Sustainable Development Goals Series

The **Sustainable Development Goals Series** is Springer Nature's inaugural cross-imprint book series that addresses and supports the United Nations' seventeen Sustainable Development Goals. The series fosters comprehensive research focused on these global targets and endeavours to address some of society's greatest grand challenges. The SDGs are inherently multidisciplinary, and they bring people working across different fields together and working towards a common goal. In this spirit, the Sustainable Development Goals series is the first at Springer Nature to publish books under both the Springer and Palgrave Macmillan imprints, bringing the strengths of our imprints together.

The Sustainable Development Goals Series is organized into eighteen subseries: one subseries based around each of the seventeen respective Sustainable Development Goals, and an eighteenth subseries, "Connecting the Goals", which serves as a home for volumes addressing multiple goals or studying the SDGs as a whole. Each subseries is guided by an expert Subseries Advisor with years or decades of experience studying and addressing core components of their respective Goal.

The SDG Series has a remit as broad as the SDGs themselves, and contributions are welcome from scientists, academics, policymakers, and researchers working in fields related to any of the seventeen goals. If you are interested in contributing a monograph or curated volume to the series, please contact the Publishers: Zachary Romano [Springer; zachary.romano@springer.com] and Rachael Ballard [Palgrave Macmillan; rachael.ballard@palgrave.com].

Phong Tan Nguyen

Nature-based Solutions in Mangrove-mud Coast Protection

From Policy to Practice in the Vietnamese Mekong Delta

Springer

Phong Tan Nguyen
Faculty of Environment and Labour Safety
Ton Duc Thang University
Ho Chi Minh City, Vietnam

College of Science and Engineering
James Cook University
Townsville, QLD, Australia

ISSN 2523-3084 ISSN 2523-3092 (electronic)
Sustainable Development Goals Series
ISBN 978-981-95-9940-0 ISBN 978-981-95-9941-7 (eBook)
https://doi.org/10.1007/978-981-95-9941-7

This Springer imprint is published by the registered company Springer Nature Singapore Pte Ltd.
The registered company address is: 152 Beach Road, #21-01/04 Gateway East, Singapore 189721, Singapore

For my family and my children

Preface

This book discusses the integration, development and use of nature-based solutions (NbS) to protect mangrove-mud coasts in the Vietnamese Mekong Delta (VMD). In short, it examines the gaps in the use of such solutions to control coastal erosion in the existing legislative document and practice in the region. Although natural processes are intensively emphasized as an integral part of coastal protection in the legislative documents, they have not been adequately considered when designing and implementing the existing coastal erosion control measures. These measures have also had limited success in controlling erosion. By contrast, local communities have mimicked natural processes to develop NbS that have successfully controlled coastal erosion, promoted the natural growth of mangrove species, and facilitated sediment accumulation. These NbS have been widely reported in the literature. Therefore, this book confirms the feasibility and effectiveness of using local knowledge and local communities to implement such solutions. Additionally, the book's technical descriptions and illustrations of coastal erosion will serve as a reference for future erosion control initiatives.

Why is another book on nature-based solutions for controlling coastal erosion needed? Many books and reviews on this topic have already been published. In fact, a review of the existing scientific literature reveals that research on these solutions is advancing rapidly. Most of the existing literature reports on the various NbS developed by researchers and technical staff working for government agencies. However, given the predicted negative impacts of climate change, rising sea levels, and coastal erosion on mangrove-mud coasts, there is an urgent need to develop and implement cost-effective erosion control measures, and time is running out to develop and test a comprehensive, single NbS. As the causes of coastal erosion vary from place to place, developing NbS requires an in-depth understanding of not only natural processes and their application in the local context. Meanwhile, local communities have developed their own knowledge of natural processes based on their daily observations and experiences in their areas. Although NbS are important and are being promoted globally, there is a lack of detailed documentation on how local communities have used their knowledge of natural processes for developing these solutions. Therefore, local communities should be considered an integral part of coastal erosion control measures, with their knowledge prioritized and incorporated into the design and implementation of such measures. This book will achieve its goal if it increases awareness and appreciation among coastal managers and other relevant parties of the natural processes involved, the NbS development by local communities and the roles and contributions of local communities in this process.

Ho Chi Minh City, Vietnam — Phong Tan Nguyen

Acknowledgments

This book would not have been possible without the support of the many kind people around me. Special thanks go to Dang Quang Huy and Gian Van Toan, who provided invaluable advice and information on coastal protection structures and mangrove forests, and to the local community members in Tan Phu Dong district who shared their knowledge of coastal protection measures in the area. I would also like to thank Huynh Phuoc Hai for facilitating my field trips to Tien Giang province, Nguyen Thanh Tung for his expert guidance on coastal protection measures, and Tong Van Anh, Nguyen Van Thua, Nguyen Phi Thong, Nguyen Thanh Trung, Nguyen Van Loi, Nguyen Quang Huy and Nguyen Quang Hoang for their unwavering support. Finally, I would like to express my gratitude to Nguyen Van Thua and Nguyen Thi Minh Huong for their beautiful photographs.

I would also like to thank Dr. Huynh Van Quoc and Nguyen Ngoc Tien for their advisory comments and advice, James Cook University and Ton Duc Thang University for administrative support and library services, and the GTZ project management board for providing me with a unique opportunity to work with Vam Ray community to develop NbS to coastal erosion control in Vam Ray, Hon Dat, and Kien Giang.

I would especially like to thank the communities in Tien Giang, Ben Tre, Soc Trang, Tra Vinh, Bac Lieu, Ca Mau, and Kien Giang for welcoming me and inviting me into their homes for meals, and helping me gather information. I would especially like to thank my parents and my two children for their support, patience, and encouragement.

Last but certainly not least, I would like to thank my wife, Nguyen Thi Minh Huong, for her sacrifices and support during the time I wrote this book. Without her, I could not have finished this book.

This publication provides detailed documentation of the author's experiences, alongside a critical review of the existing scientific literature. The views expressed herein do not necessarily reflect those of Ton Duc Thang University or James Cook University, nor do they criticize the management of the VMD coast or the efforts of those involved in its management and use.

Competing Interests The author has no competing interests to declare that are relevant to the content of this manuscript.

Contents

List of Figures

List of Tables

About the Author

Phong Tan Nguyen, MAppSc, PhD (James Cook University, Queensland, Australia) is a conservation practitioner who has worked in various environments, including terrestrial, wetland, coastal, and island in developing countries of Asia for more than 25 years. He completed the Master of Applied Science program in Protected Area Management at James Cook University in 2012. In 2016, he completed his PhD program at James Cook University. His PhD research focused on identifying mechanisms to integrate local and scientific knowledge for managing sustainably the muddy coasts of Kien Giang, Vietnam and Brebes Regency, Indonesia. He completed his postdoctoral research program at James Cook University in June 2019.

To date, he has published more than 36 ISI papers on developing and using NbS to control coastal erosion in the Vietnamese Mekong Delta and Brebes Regency, Indonesia. He is currently an adjunct lecturer at College of Science and Engineering, James Cook University, Queensland, Australia, as well as a research fellow at the Faculty of Environment and Labour Safety, Ton Duc Thang University, Ho Chi Minh City, Vietnam.

His research currently focuses on sustainable natural resource management, controlling coastal erosion using NbS, community capacity building, and policy advocacy and analysis.

List of Abbreviations

DARD	Department of Agriculture & Rural Development
GoV	Government of Vietnam
MARD	Ministry of Agriculture & Rural Development
NbS	Nature-Based Solutions
PPC	Provincial People's Committee
VMD	Vietnamese Mekong Delta
VNA	Vietnam National Assembly
VPM	Vietnam Prime Minister

Introduction to the Vietnamese Mekong Delta Coast

1

Abstract

This chapter provides an overview of the administrative and geographical features of the VMD coast, including its location, total area, hydrology, tidal regimes, and sedimentation. It also presents the current management and use of the VMD coast, including sustainable management and use of coastal mangrove forests for socio-economic development and protection the VMD coast using of shoreline and offshore structures to mitigate negative impact of climate change and rising sea levels. This introduction provides readers with the basic information needed to understand the context of the region and the coastal changes it has experienced in recent years.

Keywords

Coastal mangroves · Offshore breakwaters · Protection forests · Sea dykes · Sluice gates · Rising sea levels

1.1 Introduction

VMD is located in the southwest part of Vietnam, to the southwest of Ho Chi Minh City. It borders the East Sea of Vietnam to the east, the Gulf of Thailand to the west, and Cambodia to the north and northeast. Spanning 12 provinces, and one centrally runs city: Can Tho City, the delta covers 4.0 million hectares, accounting for over 13% of Vietnam's total land area. Home to approximately 17 million people, the delta makes up about 18% of the country's population (Vietnam's Data Service Portal 2023). After July 1, 2025, GoV merged all provinces, leaving VMD with four provinces and one centrally run city.

The region is influenced by Asia's tropical monsoons, including the southwest and northeast monsoons. The southwest monsoon brings moisture from the sea to the mainland and occurs from May to September. The northeast monsoon occurs from November to March. The east coast experiences irregular semi-diurnal tides, with tidal amplitudes ranging from 3.0 to 3.5 m, while the west coast experiences irregular diurnal tides with amplitudes ranging from 0.8 to 1.2 m.

The region is dominated by three major river systems. In the east, the Tien and Hau rivers flow into the East Sea through nine estuaries: the Tieu, the Dai, the Ba Lai, the Ham Luong, the Co Chien, the Cung Hau, the Dinh An, the Ba Thac, and Tran De. Due to sedimentation, the Dinh An and Tran De rivers have narrowed, resulting in the formation of two small canals: the Con Coc river flows into the Dinh An river estuary branch, and the Con Tron river flows into the Tran De river estuary branch. In the west is Cai Lon-Cai Be river system, which discharges into the Gulf of Thailand (Fig. 1.1).

P. T. Nguyen, *Nature-based Solutions in Mangrove-mud Coast Protection*, Sustainable Development Goals Series, https://doi.org/10.1007/978-981-95-9941-7_1

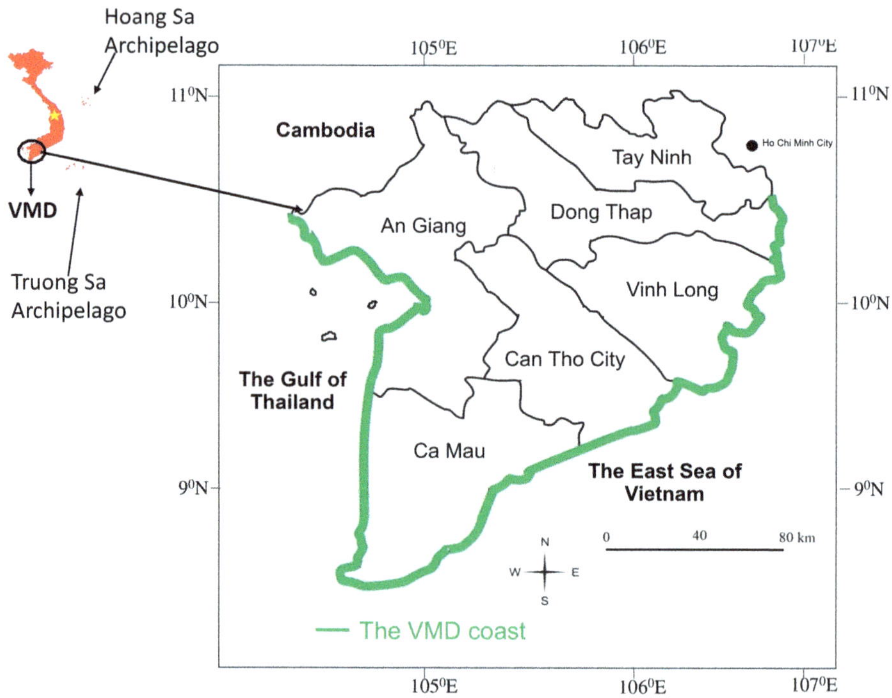

Fig. 1.1 The VMD coast and its location in Vietnam. (Adapted from Nguyen and Loi (2023))

The evolution of the VMD coast is strongly influenced by the discharge regimes of the Mekong River (the Tien and Hau rivers) and the Cai Lon and Ca Be rivers and a diverse network of natural and man-made channels. The Tien and Hau rivers discharge floodwater and sediments into the East Sea, while the Cai Lon and Cai Be rivers transport them into the West Sea. Natural channels in the region help to retain freshwater for agricultural production, such as rice, cash crops, and vegetables during the rainy season. They also transport floodwater and sediments to the seas. During the 1970s, channels were dug across the region primarily to remove acid sulphate from the soil and make it suitable for agriculture as well as to facilitate local boating.

Tidal currents control sediment transport in the east, while wave action drives it in the West Sea. Tidal waves propagate from the northeast to the southwest along the East Sea, transporting the sediments along the shoreline and forming a significant longshore sediment component (Liu et al. 2017). These currents intensify during the northeast monsoon season. Coarse-grained sediment tends to be deposited in and around estuarine areas. Fine-grained sediment is transported along the shoreline during the rainy season and is washed ashore by tidal waves (Eidam et al. 2017; Xing et al. 2017). The Mekong River has a high sediment discharge of approximately amount of 110–160 Mt/year, resulting in a fast seaward progradation rate of up to 30 m/year for the subaerial delta plain. The river produces deposits with a significant offshore sediment accumulation rate of 1–10 cm/year (Liu et al. 2017).

1.2 The VMD Coast and Its Current Management

VMD has a coastline of approximately 800 km (Fig. 1.1 and Table 1.1), which is dominated by coastal mangrove forests and various types of land. The GoV is currently managing the VMD coast by protecting coastal mangrove forests and

Table 1.1 Mangrove forests by province of VMD (Institute for Forestry Planning and Inventory 2015)

No.	Province (new province after merging—July 1, 2025)	Coastline (km)	Mangrove belt in length (km)	Total area (ha)	Total forest area (ha)	Total barren area/ mudflats (ha)
1	Tien Giang (Dong Thap)	32	21	4350	4282	68
2	Ben Tre (Vinh Long)	65	65	4580	1814	2766
3	Tra Vinh (Vinh Long)	65	65	11,350	8000	3350
4	Soc Trang (Can Tho City)	72	72	13,000	8476	4524
5	Bac Lieu (Ca Mau)	56	56	6450	4149	2301
6	Ca Mau (Ca Mau)	310	97	18,180	9571	8609
7	Kien Giang (An Giang)	209	169	23,000	9400	13,600

constructing shoreline and offshore structures to prevent further erosion.

1.2.1 The Protection of Coastal Mangrove Protection Forests

During the Vietnam-Cambodia War in 1978, natural mangrove forests in seven coastal provinces were cleared to make way for the construction of earthen surface roads that were used to transport ammunition and weapons for the war (Nguyen et al. 2017). Therefore, these roads are known locally as "national defense sea dykes." In the 1980s, GoV was committed to creating 1 million hectares of new forests (both terrestrial and coastal) to offset the loss of forests. To achieve this commitment, GoV issued and implemented *the 661-Afforestation Program*. The *Rhizophora* species was selected for the afforestation program in coastal areas for two reasons. Firstly, this mangrove species was considered a "real" mangrove forest at that time. Secondly, fully grown *Rhizophora* trees could be harvested for wood timber or charcoal production. Terrestrial afforestation techniques were used in coastal areas. These techniques involved two steps. As the first step, the vegetation cover of designated areas was cleared prior to afforestation. The second step involved planting of propagules in rows with high density. Propagules of *Rhizophora* species were collected and transported by boat from the Minh Hai Plantation (now Ca Mau province) to afforestation sites. This type of planting was employed for a long period of time in the region, resulting in stands of pure *Rhizophora* species along most of the coast (Nguyen et al. 2017; Nguyen 2024).

In 1991, Vietnam passed the *Law on Forest Protection and Development Law,* which was subsequently revised in 2017 (VNA 2017). Under this legislation, both terrestrial and coastal forests are designated as protected areas. Human activities (e.g., collection of forest and non-forest products, wood collection, etc.) without permission from the relevant government agencies are prohibited in these areas. To implement this law, seven coastal provinces designated coastal mangrove forests as protected areas and assigned DARD to manage these protected areas in each province. At provincial level, DARD established a board that helped DARD manage these forests, with the permission from the provincial level People's Committee. These protected areas formed as a natural protective belt that helped protect the VMD coast from negative effects of climate change and rising sea levels (VPM 2009).

From 2001 to 2009, GoV issued numerous legislative documents that encouraged the sustainable management and use of mangrove forests in VMD for socio-economic development. These legislative documents involved included designation of coastal mangrove forests into national parks and nature reserves in accordance with the Vietnamese Law on Forestry (VNA 2017) as well as the use of coastal mangrove protection forests and coastal/marine environments for socio-economic development and coastal protection (VPM 2015). Since 2009, seven coastal provinces have upgraded or replaced earthen sea dykes with concrete sea dykes and sluice gates to better to adapt to climate change and rising sea levels (VPM 2009).

For management purposes, the coastal mangrove protection forests were categorized into three ecological mangrove zones in accordance with their ecology, dynamics, and occurrence. The zones include the primary mangrove belt, the secondary mangrove belt, and the intertidal mudflats and eroding sites. The primary mangrove belt is located between intertidal mudflats and naturally growing *Avicennia* trees. This belt constantly moves and gradually extends toward the sea alongside the mudflats. The secondary mangrove belt is a stable and defined coastal mangrove forest that extends from the sea dike and/or rural road to the sea. It is located adjacent to the primary mangrove belt. Intertidal mudflats and/or eroding areas serve as buffer zones for the primary mangrove belt (Kien Giang PPC 2011) (Table 1.1).

1.2.2 The Construction of Shoreline and Offshore Structures to Protect the VMD Coast

In addition to protecting existing coastal mangrove forests, GoV has been constructing a complex system of sea dykes along the VMD coast since 2009. Earthen sea dykes were originally constructed to store freshwater for agricultural irrigation and for military or national defense purposes, particularly during the 1978 war (The "Southeast War" on the Vietnam-Cambodia border). After the war, the crest of the earth sea dykes was levelled to create rural transport routes, known locally 'rural roads'. Since 2009, these earthen sea dikes have been upgraded or replaced with concrete dikes, and sluice gates have been constructed in seven provinces to better adapt to climate change and rising sea levels (VPM 2009). By 2017, the government had invested over VND 4400 trillion in improving mangrove management and coastal protection (MARD 2017). By 2025, a complex system of sea dikes and sluice gates had been built along the VMD coast, with more offshore pile-rock breakwaters constructed in Tien Giang (Tu et al. 2022a, b) and Ca Mau (Luom et al. 2021) and Kien Giang (Table 1.2).

1.3 Conclusions

The VMD coast has been shaped by two different tidal regimes, a diverse river system, coastal mangrove forests, and sedimentation. Because the VMD coast plays a crucial role in Vietnam's socioeconomic development, GoV has empha-

Table 1.2 The offshore structures along the VMD coast as in February 2023

Province	Concrete sea dyke (km)	Geo-tube (km)	Pile-rock breakwater (km)	Hollow triangular breakwater (km)	Hollow cylinder breakwater (km)
Tien Giang	11	2	16	4	0
Ben Tre	27	0	0	0	0
Tra Vinh	59	0	0	0	0
Soc Trăng	38	0	1.5	0	0
Bac Lieu	21	0	4.5	0	0
Ca Mau	38	0	145	0	0.18
Kien Giang	47	0	64	0	0

sized its efforts to sustainably manage, protect as well as use this region for socio-economic development and for better adaptation to negative impacts of rising sea levels and climate change. These efforts included protection of coastal mangrove protection forests into protected areas and construction of shoreline and offshore structures to protect the VMD coast from erosion.

References

Eidam EF, Nittrouer CA, Ogston AS, DeMaster DJ, Liu JP, Nguyen TT, Nguyen TN (2017) Dynamic controls on shallow clinoform geometry: Mekong Delta, Vietnam. Cont Shelf Res 147:165–181. https://doi.org/10.1016/j.csr.2017.06.001

Kien Giang PPC (2011) The decision on issuing the regulations of mangrove planting, protection, and utilization in Kien Giang province. Kien Giang PPC, Rach Gia, Kien Giang

Liu JP, DeMaster DJ, Nittrouer CA, Eidam EF, Nguyen TT (2017) A seismic study of the Mekong subaqueous delta: proximal versus distal sediment accumulation. Cont Shelf Res 147:197–212. https://doi.org/10.1016/j.csr.2017.07.009

Luom TT, Nguyen TP, Anh NT, Tung NT, Tu LX (2021) Using fine-grained sediment and wave attenuation as a new measure for evaluating the efficacy of offshore breakwaters in stabilizing an eroded muddy coast: insights from Ca Mau, The Mekong Delta of Vietnam. Sustainability 13(9):4798. https://doi.org/10.3390/su13094798

MARD (2017) Coastal mangrove forests in the Mekong Delta decreased in area over the past 15 years. Available online at: http://vea.gov.vn/vn/quanlymt/baotondadangsh/tintuchoatdong/tintucvasukien/tintuc/Pages/Di%E1%BB%87n-t%C3%ADch-r%E1%BB%ABng-ng%E1%BA%ADp-m%E1%BA%B7n-%C4%90%E1%BB%93ng-b%E1%BA%B1ng-s%C3%B4ng-C%E1%BB%ADu-Long-gi%E1%BA%A3m-trong-5-n%C4%83m-qua.aspx. Accessed 12 Dec 2018

Nguyen TP, Loi LT (2023) Extent of reference to science and the efficacy of coastal management policies: insights from the Vietnamese mekong delta. J Coast Conserv 27–55. https://doi.org/10.1007/s11852-023-00983-7

Nguyen TP (2024) Erosion of mangrove dominated muddy coasts in Vietnam and Indonesia: processes and accelerating factors. Reg Stud Mar Sci 79:103821. https://doi.org/10.1016/j.rsma.2024.103821

Nguyen TP, Cottrell A, Parnell KE (2017) Human activities and coastal erosion on the Kien Giang coast, Vietnam. J Coast Conserv 21(6):967–979. https://doi.org/10.1007/s11852-017-0566-9

Tu LX, Phong NC, Thanh VQ, Quan TQ, Wright DP, Anh DT (2022a) Multi-scale modelling for hydrodynamic and morphological changes of breakwater in coastal Mekong Delta in Vietnam. J Coast Conserv 26:18. https://doi.org/10.1007/s11852-022-00866-3

Tu LX, Manh LH, Ba HT, Duong DV, Vu HTD, Wright D, Bui VH, Anh DT (2022b) Wave energy dissipation through a hollow triangle breakwater on the coastal Mekong Delta. Ocean Eng 245:110419. https://doi.org/10.1016/j.oceaneng.2021.110419

Vietnam's Data Service Portal (2023) Area, population, and population density of Vietnam in 2023. Vietnam's Data Service Portal, Hanoi

VNA (2017) Law on Forestry Vietnam. VNA, Hanoi

VPM (2009) Program to upgrading and reinforcing the sea dyke system from Quang Ngai to Kien Giang provinces. VPM, Hanoi

VPM (2015) Regulations on the management of protection forests. VNA, Hanoi

Xing F, Meselhe EA, Allison MA, Weathers HD III (2017) Analysis and numerical modeling of the flow and sand dynamics in the lower Song Hau channel, Mekong Delta. Cont Shelf Res 147:62–77. https://doi.org/10.1016/j.csr.2017.08.003

2 Coastal Erosion Along the VMD Coast

Abstract

This chapter illustrates the four most common erosion processes that prevailed along the VMD coast. These were (a) eroding areas with remnants of pond dikes, (b) eroding areas with vertically collapsed blocks, (c) eroding areas with exposed agricultural areas, and (d) eroding areas with the construction of sea dikes and sluice gates. These processes were illustrated and summarized using photographs taken during field trips, the information obtained from semi-structured interviews regarding the management and use of these sites. The illustrations and summaries demonstrate the roles of natural and anthropogenic factors in contributing to coastal erosion. Anthropogenic factors included policy-based activities such as clearing natural mangrove vegetation to make way for planting and the excessive use of coastal mangrove forests to expand development areas beyond permitted limits. These factors have fragmented and exposed existing mangrove forests to erosion. Natural factors, such as strong waves and seasonal currents, repeatedly penetrated these fragmented, vulnerable areas, eventually causing them to collapse and uproot mature *Rhizophora* trees.

Keywords

Incident waves · Uprooted trees · Vertical collapse · Dyke remnants · Local boating

2.1 Introduction

By 2015, the VMD coast had been severely eroded (Anthony et al. 2015; Phan et al. 2015), with 268 km (approximately 35% of the total coastline) having been affected (MARD 2017). The maximum annual erosion rate was approximately 45 m, equivalent to the loss of 15,000 ha of mangroves between 2011 and 2016 (MARD 2017). Many aquaculture areas were abandoned due to erosion and became unsuitable for aquaculture (Nguyen 2024) (Fig. 2.1).

Recent studies have examined the status of coastal erosion, its causes, and its consequences along the VMD coast. For instance, researchers used aerial photography and specialized software to examine shoreline changes in the VMD coast (Anthony et al. 2015; Phan et al. 2015; Besset et al. 2019). Other studies have investigated the causes of erosion along the VMD coast, including both human-made and natural disasters (Anthony et al. 2015; Nguyen et al. 2017; Nguyen

P. T. Nguyen, *Nature-based Solutions in Mangrove-mud Coast Protection*, Sustainable Development Goals Series, https://doi.org/10.1007/978-981-95-9941-7_2

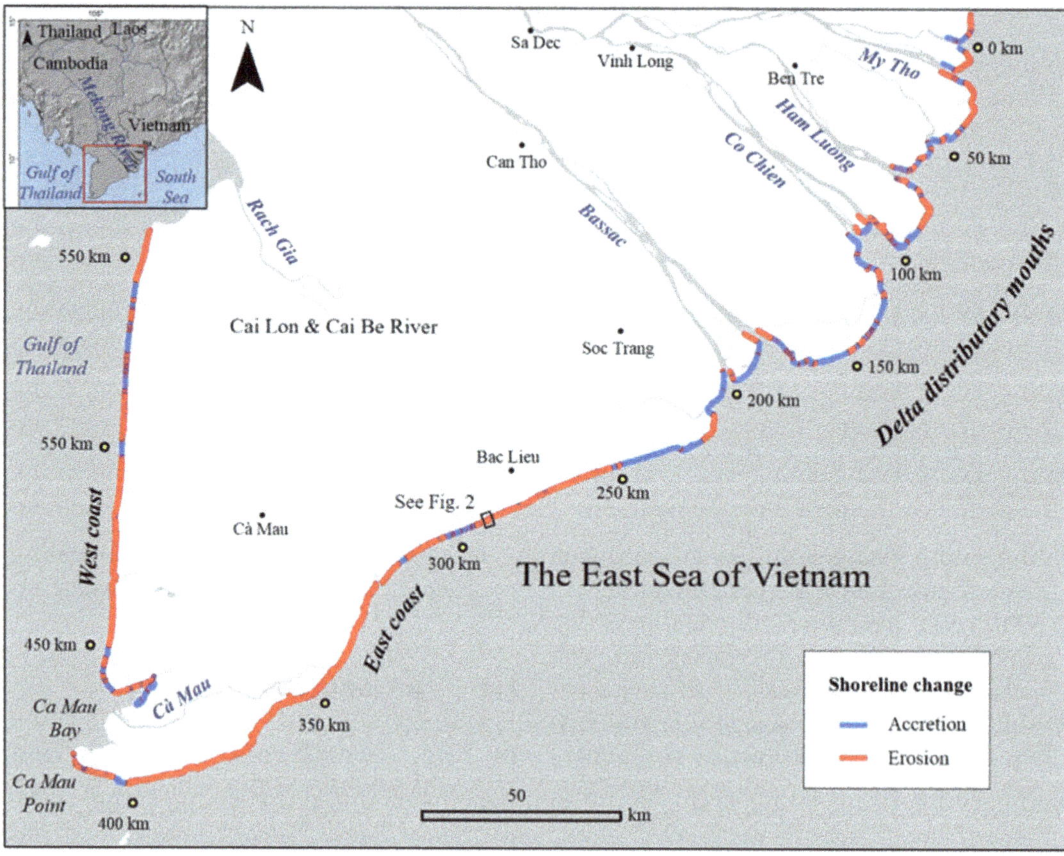

Fig. 2.1 The VMD coast and coastal erosion. (Adapted from Besset et al. (2019))

and Parnell 2019; Nguyen 2024). Human activities contributing to erosion included the misuse of forests and protected land for aquaculture, the improper design of sea dykes and sluice gates, illegal and legal exploitation of sand mining, and the consequences of hydropower dam construction on the upper Mekong River. Natural factors included changes in sediment flow (Oanh et al. 2002; Oanh et al. 2021), rising sea levels, and climate change (MARD 2017; Vietnam MonRE 2020).

Despite the abundance of scientific knowledge about coastal erosion in the region, a clear picture of the extent of coastal erosion remains elusive. In some cases, coastal communities and local governments are unaware of the negative effects of their actions on the VMD coast. Therefore, a detailed account of coastal erosion is crucial for the effective management of the VMD coast and as a foundation for further research. This chapter provides illustrations and summaries of the coastal erosion that occurred along VMD. This illustration was based on the author's experience of years working in the area, detailed observations and documentation from recent field visits to the sites, analysis of grey literature archived in the offices of DARD of each province, and a critical review of peer-reviewed publications. Some of these data and the findings were previously published (Nguyen 2024).

Grey literature was collected during the field visits and semi-structured interviews with local communities and technical staff working for

DARDs. A critical review was undertaken using two major domains: *SpringerLink* and *ScienceDirect*. Key words used for searching for and downloading articles included "*Mekong Delta coast*," "*coastal erosion in Mekong Delta coast; eroding Mekong coast*." All hits were classified and analyzed using thematic analysis, to avoid duplication (Table 2.1).

Table 2.1 Summary of grey and peer-reviewed literature retrieved from the main domains

No.	Documents (year)/articles (journal)	Department/ author(s)
Grey literature		
1	Report on evaluating climate change and weather forecast in Tien Giang (2020)/Progress report	DARD Tien Giang
2	Report on master plan for development of agro-forestry and aquaculture in Tien Giang until 2020, with vision up to 2030 (2017)/Progress report	Tien Giang PPC
3	Report on coastal mangrove protection forest (2021)/Progress report	DARD Tien Giang
4	Report on coastal mangrove protection forest (2024)/Progress report	DARD Tien Giang
5	Report on preliminary summary of the implementation of forest allocation in Tra Vinh (2021)/Progress report	Tra Vinh PPC
6	Report on preliminary summary of the implementation of forest allocation in Ca Mau (2021a)/Progress report	DARD Ca Mau
7	Statistics of coastal and riverbank erosion in Ca Mau (2024)/Statistics report	DARD Ca Mau
8	Review of forest and forest lands in Ca Mau (2021b)/Progress report	DARD Ca Mau
9	Report to the Vietnamese Central Office Working Group on surveying dikes and checking the landslide situation in Ca Mau (2024)/Progress report	Ca Mau PPC
10	Report on the implementation of five year of Decision 51 by Kien Giang PPC on afforestation and forest protection in Kien Giang (2011)/Progress report	Kien Giang PPC
11	Project of coastal mangrove protection forest project and afforestation between 2012 and 2016 in Kien Giang, Vietnam/Project document	DARD Kien Giang
Peer-reviewed literature		
12	Coastal land use planning in Ben Tre: constraints and recommendations	Nguyen and Parnell (2019)
13	Mangroves and shoreline erosion in the Mekong River delta, Viet Nam	Besset et al. (2019)
14	Coastal mangrove squeeze in the Mekong Delta	Phan et al. (2015)
15	Erosion of mangrove dominated muddy coasts in Vietnam and Indonesia: processes and accelerating factors	Nguyen (2024)
16	Human activities and coastal erosion on the Kien Giang coast, Vietnam	Nguyen et al. (2017)
17	Linking rapid erosion of the Mekong River delta to human activities	Anthony et al. (2015)
18	Drivers of coastal shoreline change: case study of Hon Dat coast, Kien Giang, Vietnam	Hoa et al. (2015)
19	Mechanism of erosion zone formation based on hydrodynamic factor analysis in the Mekong Delta coast, Vietnam	Hanh and Bui (2023)
20	Coastal erosion trend analysis using a combination of remote sensing and hydrodynamic models: case study of Ca Mau Cape, Mekong Delta	Bui and Hanh (2022)
21	Erosion of mangrove dominated muddy coasts in Vietnam and Indonesia: processes and accelerating factors	Nguyen (2015)
22	Erosion of the coastal Mekong delta: assessing natural against man induced processes	Marchesiello et al. (2019)

The Google Earth software was used to locate erosion sites that were identified from the review. This resulted in images showing the condition of the erosion sites and mangrove forests along the VMD coast. The literature review and the analysis of the Google Earth imagery led to the development of open-ended questions, which were subsequently used in semi-structured interviews. The aim of these interviews was to understand the causes of coastal erosion and the measures taken to stop it.

Multiple field visits were organized along the VMD coast to verify erosion sites and locate additional coastal erosion sites between May 2015 and October 2025. During these visits, the author documented the status of these sites, took photographs, and conduct semi-structured interviews with relevant stakeholders. Open-ended questions were first tested in an interview with a technical staff member working for DARD Tien Giang. These questions were then revised for subsequent interviews (Table 2.2).

Fourteen semi-structured interviews were conducted with fourteen technical staff of DARD office in the seven provinces of VMD. Two technical staff from each DARD were interviewed at their office. Each interview lasted one (1) hour. Seven focus group interviews (120 interviewees in total) were also conducted with the support of the seven DARD offices of seven coastal provinces in VMD. Three criteria were used to select focus group members: (a) individuals involved in the coastal protection activities such as mangrove allocation, mangrove afforestation, and coastal protection, (b) and individuals whose land or ponds had been eroded, and (c) individuals extensively involved in coastal tourism development. Purposive sampling (Neuman 2011) was used for these focus group interviews as participants with knowledge of mangrove forests, mangrove-based aquaculture, and muddy coasts were required for this study. Before each group interviews, group members were provided with the information relating the objectives of the study and interview, including their rights and the right to refuse the interview. Group members gave verbal consent before the interviews. Each group interview was organized in a local location selected by the groups in the community and lasted approximately one (1) hour. Maps, photographs taken during the field observations and the findings from the critical review were used to stimulate the discussion about the causes of erosion and the erosion processes. The data and findings from field observations, photographs, and semi-structured interviews were analyzed and categorized into diagram of drivers contributing to erosion using participatory diagramming (Neuman 2011).

Table 2.2 The open-ended questions used in semi--structured interviews in VMD

Open-ended questions (English translation)
1. What did the area look like in the past?
2. When did coastal erosion occur?
3. How did coastal erosion begin?
4. What activities are permitted in coastal mangrove protection forests?
5. What measures have been undertaken to control erosion?
6. What do you think about the relationship between coastal erosion and socio-economic development (aquaculture ponds) and agriculture production?

2.2 Coastal Erosion Along the VMD Coast

The VMD coast has experienced four types of coastal erosion: (a) eroding areas with remnants of pond dikes, (b) eroding areas with vertically collapsed blocks, (c) eroding areas with exposed agricultural areas, and (d) eroding areas with construction of sea dykes and sluice gates.

Fig. 2.2 Eroding areas with remnants of pond dykes and sluice gates in An Minh, An Giang, full of remnants of pond dykes. (**1**) Google Maps shows breached pond dykes. (**2**) an eroding pond. (**3**) manually constructed pond gate. (**4**) eroded coastal areas with remnants of pond dykes visible during low tide. (**a**) pond dyke with mangrove trees, mainly of the *Rhizophora* species, (**b**) pond area, (**c**) pond area breached and eroded without mangrove protection, (**d**) remnants of pond dykes, (**e**) manually constructed pond gates

2.2.1 Eroding Areas with Remnants of Pond Dykes

- *Locations of These Areas in VMD*

These eroding areas consisted of remnants of pond dykes and dead *Rhizophora* trees. These areas were found in both privately owned and government-managed coastal areas on the eastern and western sides of the VMD coast (Fig. 2.2).

- *Historical Management and Use of These Eroding Areas*

The semi-structured interviews and the focus group interviews revealed that these areas were once designated for mangrove protection and aquaculture development, with the aim of improving local livelihoods. These designations were made through allocation contracts signed by the management board of the coastal mangrove protection forests (hereafter referred to as the contractor) and the local communities living around or near these forests (hereafter referred to as the contractees). An allocation contract lasted for 49 years. In return for the use of mangrove forests for aquaculture development, contractees were required to be involved in mangrove protection activities, including mangrove afforestation and the protection of existing mangrove forests (please refer to Chap. 3 for further information on the allocation program) (Nguyen et al. 2017).

However, the contracts did not include any instructions regarding the use of mangrove protection forests for aquaculture development. For example, there were no guidelines on where aquaculture ponds and mangrove plantings should be located within the allocated areas. Contractees used their own knowledge to construct the ponds. They cleared the allocated mangrove forests to make space to construct ponds and gates, which were surrounded by manually constructed dykes. Contractees planted *Rhizophora* seedlings on the pond dykes and counted this as mangrove afforestation towards the required percentage of mangrove protection (Nguyen et al. 2017).

In many cases, however, mangrove forests were illegally cleared to create larger ponds than permitted. Due to the demand for water exchange for pond operation, contractees intended to build ponds in mangrove forests growing on intertidal mudflats or primary mangrove belts. This left narrow protective barriers (Nguyen et al. 2017). This uncontrolled clearance contributed to the fragmentation of protected mangrove forests. Repeated penetration by strong incident waves from the northeasterly wind regime operating from November to April further fragmented the thin mangrove forests. Over time, the ponds were abandoned, and the mature *Rhizophora* trees were uprooted (Nguyen and Parnell 2019).

The field visits revealed that remnants of the pond dykes could still be seen along VMD at low tide. These eroded areas on the western side were inundated with remnants of the pond dykes and dead *Rhizophora* trees (Fig. 2.3), while the eroded areas on the eastern side were gradually being buried by sand (Nguyen 2024) (Figs. 2.3 and 2.4).

2.2.2 Eroding Areas with Blocks of Soil Vertically Collapsed

- *Locations of These Areas in VMD*

These actively eroding areas were the remains of soil blocks that vertically collapsed. These areas were found on both sides of the VMD coast (Fig. 2.5).

- *Historical Management and Use of These Eroding Areas*

The semi-structured interviews and the focus group interviews indicated that in the 1970s, natural mangrove forests (predominantly consisting of *Rhizophora*, *Avicennia*, and *Bruguiera* species) were cleared to make space for planting *Rhizophora* seedlings, with the aim of developing new mangrove forests. This species was chosen because local residents could harvest its wood to improve their livelihoods once the trees reached their maturity. Mature trees were harvested locally to make charcoal and fuel, as well as construction materials. The seedlings were planted at a density of 10,000 per hectare, resulting in stands of pure and dense mature *Rhizophora* trees. These areas had the same soil type: clay mixed with silt and sand. Since the 1980s, these areas have been legally protected as coastal mangrove protection forests. The high-density *Rhizophora* trees, along with the clay, silt, and sand created a favorable condition for the thriving of brackish water fiddler crabs (*Uca minax*) (Nguyen 2015; Nguyen and Parnell 2019). Over time, these crabs dug high-density burrows in these areas. During field visits, these burrows were found around the aerial roots of mature *Rhizophora* trees. Additionally, mature trees experienced a sediment deficit around their roots (Nguyen 2024).

Fig. 2.3 An eroding coastal area with remnants of pond dykes and area buried by sand in Ba Tri, Ben Tre, the eastern side of VMD. (**a**) remnants of clay soil pond dyke visible at low tide, (**b**) pond area filled with sand, (**c**) mangrove edge with mixed mangrove forests

The field visits showed that strong incident waves had penetrated and enlarged the burrows over time (Nguyen 2024). These waves also uprooted mature *Rhizophora* trees, which were subsequently removed. The penetration of these waves, particularly during the region's northeasterly wind regime from November to April, caused soil blocks to break and collapse over time. Similar processes were also observed along the eastern side of the VMD (Figs. 2.6 and 2.7).

2.2.3 Eroding Areas with Exposed Agriculture Production Area

- *Locations of These Areas in VMD*

The eroding areas were agricultural production areas completely exposed to the sea and stands of mature *Rhizophora* trees with aerial roots buried by sand. These areas were found in privately owned and government-managed coastal areas on eastern side of the VMD coast (Figs. 2.8 and 2.9).

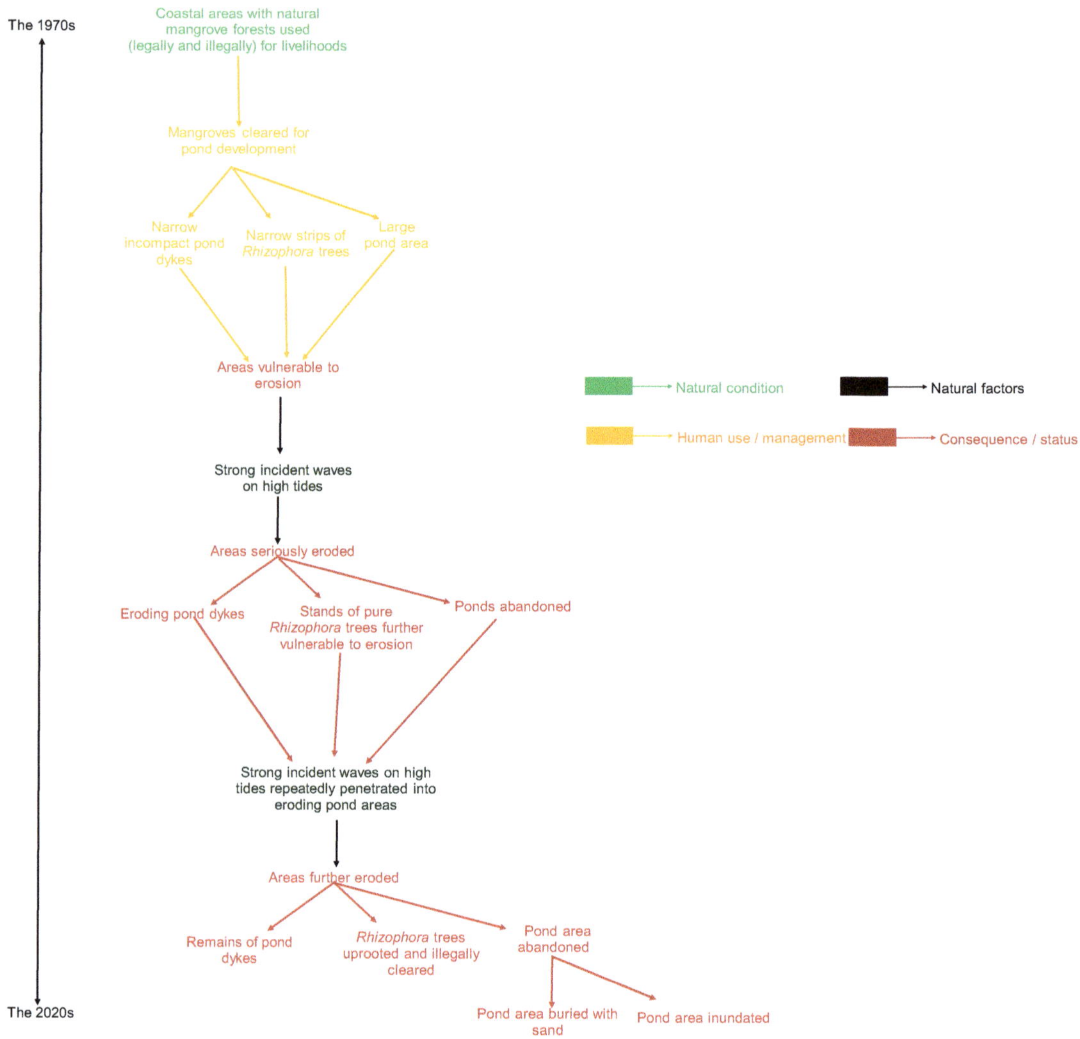

Fig. 2.4 Summary of coastal erosion and factors that significantly contributed to the erosion process in allocated mangrove forests. (Adapted from Nguyen (2024))

- *Historical Management and Use of These Areas*

These areas were either protected mangrove forests or privately owned land. Originally, these areas were natural mixed mangrove forests located approximately 350 m from the existing agricultural production areas. They were situated behind thick belts of various mangrove species. These natural mixed mangrove forests were cleared to make space for the production of cash crops, including watermelons, sweet potatoes, and cassava (Nguyen 2015). Further mangrove forests were cleared to construct ponds for culturing marine species, such as Barramundi, shrimp, and mud crabs. This left thin lines of mangrove forests facing the sea. During the 1980s, strong incident waves during high tides washed away and uprooted the remaining man-

Fig. 2.5 Coastal areas with blocks of soil vertically collapsed. (**a**) aerial roots of *Rhizophora* species are growing on the surface of the soil, (**b**) burrows were enlarged by strong incident waves over time, (**c**) blocks of soil vertically collapsed under the influence of strong incident waves

grove forests, leaving the agricultural production areas vulnerable to erosion (Fig. 2.8).

In the 1970s, areas with mature *Rhizophora* species with buried aerial roots were located approximately 350 m behind the front line of mixed natural mangrove forests. These areas are now located closer to the front line. Strong waves during high tides have uprooted the mangrove trees on the front line. At the same time, further mangrove areas were cleared to make way for expanding aquaculture ponds or agricultural production areas. This made the *Rhizophora* trees vulnerable to erosion. Gradually, the aerial roots were buried by sand which was transported ashore by strong incident waves (Fig. 2.9). The entire process is summarized in Fig. 2.10.

2.2.4 Eroding Areas with Construction of Sea Dykes and Sluice Gates

- *Locations of These Areas in VMD*

The eroding areas were sea dykes and stands of mature *Rhizophora* trees in the western side of VMD (Figs. 2.11 and 2.12).

Fig. 2.6 (**1**), (**2**), (**3**) and (**4**) shows stands of pure and dense *Rhizophora* trees because of dense planting (10,000 seedlings per ha). (**a**) deficit sediment around the aerial roots of *Rhizophora* species, (**b**) dense and pure *Rhizophora* trees growing on the soil, (**c**) uprooted pure *Rhizophora* trees, leaving the entire areas vulnerable to erosion, (**d**) blocks of soil that were collapsed by strong incident waves on high tides and have been buried with sand

- *Historical Management and Use of These Areas*

  The semi-structured interviews and the focus group interviews revealed that natural mangrove forests were cleared to build channels for local boating, as well as for shrimp and mussel farming. This process occurred in two stages. As the first stage, in the 1970s, mangrove forests were cleared to build boat channels and to divert floodwaters to the sea. Mangroves were also cleared to build earthen sea dykes that aimed at storing freshwater for rice irrigation during the dry season in VMD. In the second stage, which occurred in the 1980s, storm surges and strong incident waves, particularly during the northeasterly wind regime from November to April, eroded the intertidal mudflats, breached the aquaculture ponds and boating channels, and exposed the earthen sea dykes to erosion. Strong incident waves repeatedly broke the loosely compacted sea dykes, creating large holes. Eventually, the sea dykes were breached and collapsed. This erosion process was observed in the Hon Dat district, Kien Giang in VMD (Nguyen et al. 2017). The process can be summarized as below in Fig. 2.13.

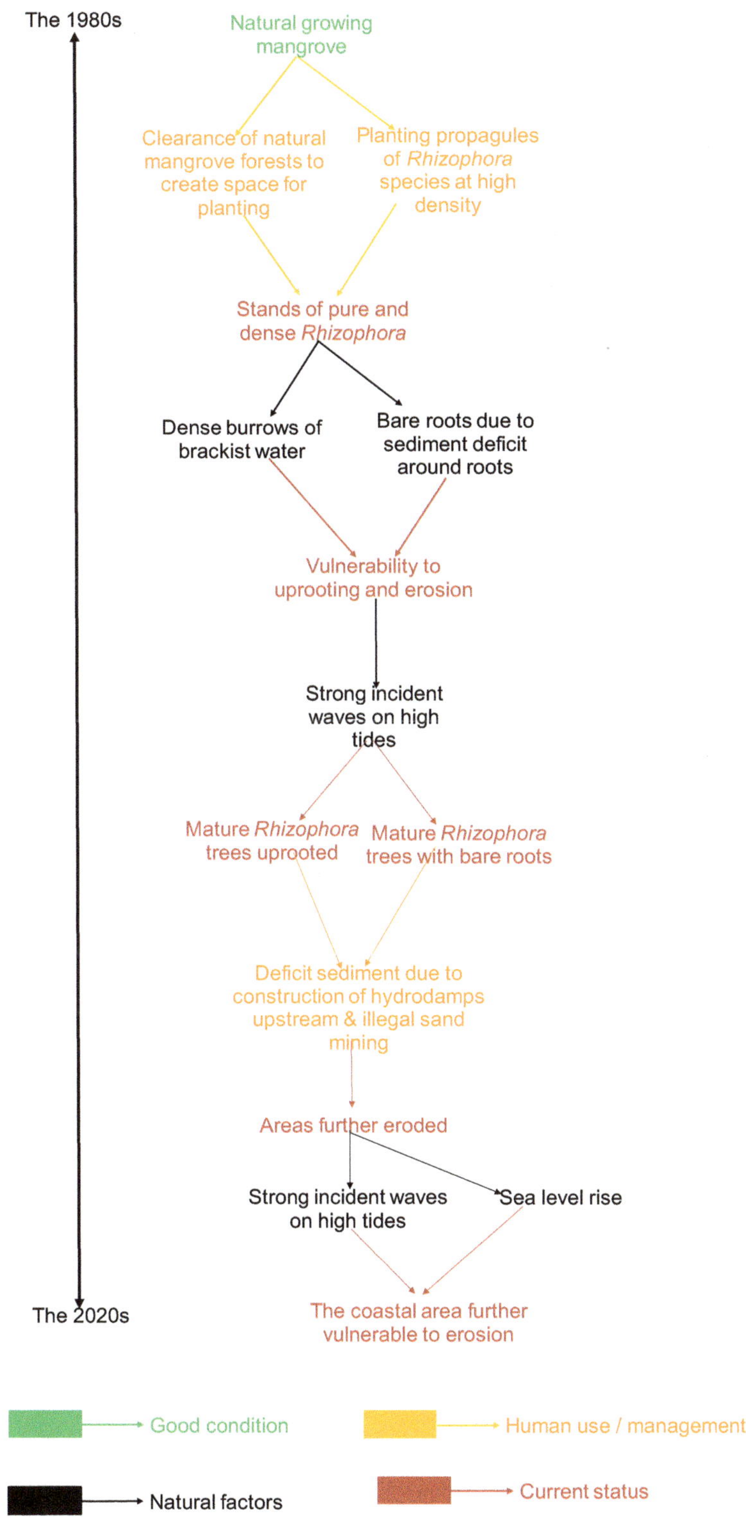

Fig. 2.7 Summary of coastal erosion and factors significantly contributing to the erosion process on stands of dense and pure *Rhizophora* species. (Adapted from Nguyen (2024))

Fig. 2.8 Mangrove areas were cleared to grow cash crops, such as watermelons, sweet potatoes, winter melons, and pumpkins. (**a**) the East Sea of Vietnam, (**b**) bunds covered with plastic for growing cash crops

2.3 Conclusions

This chapter presents the illustrations and summaries of the four most coastal erosion processes along the VMD coast, including eroding areas with remnants of pond dikes, eroding areas with vertically collapsed blocks, eroding areas with exposed agricultural areas, and eroding areas with construction of sea dykes and sluice gates. The illustrations ad summaries demonstrate the roles of human and natural factors to contributing to the coastal erosion in the region. Over time, anthropogenic activities, both legal and illegal, have fragmented and exposed coastal areas to erosion. Natural factors, including strong waves during high tides and rising sea levels, have also occurred gradually over time. These natural processes which have been accelerated by anthropogenic activities have breached vulnerable areas with collapsed blocks or remnants of pond dikes and mangrove trees along the VMD coast.

Fig. 2.9 Aerial roots of mature and standing *Rhizophora* trees were buried with sand in Tien Giang, VMD. (**a**) sand accumulation, (**b**) aerial roots of *Rhizophora* species with sediment deficit

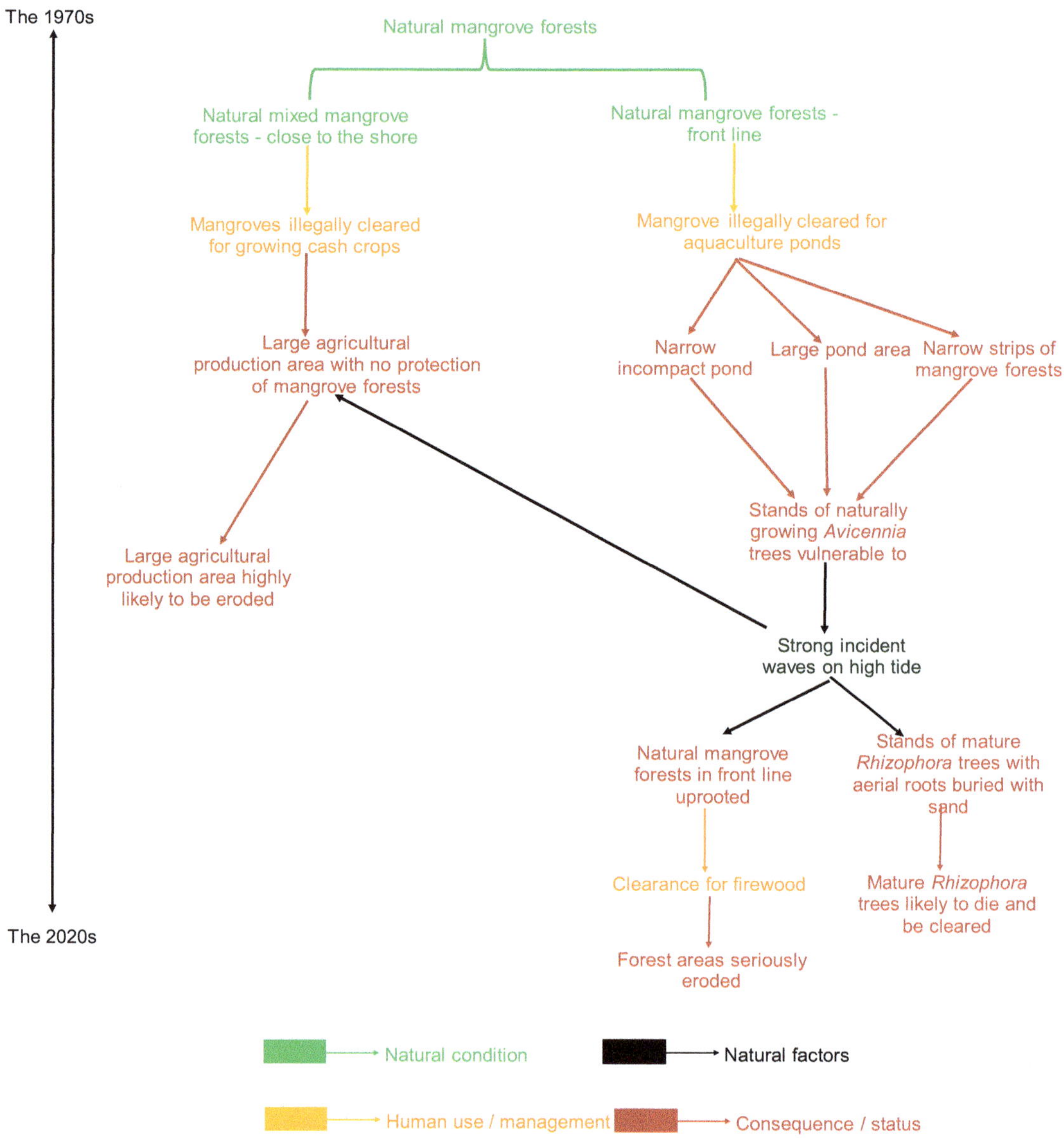

Fig. 2.10 Summary of coastal erosion with factors contributing to the erosion process in agricultural production area. (Adapted from Nguyen (2024))

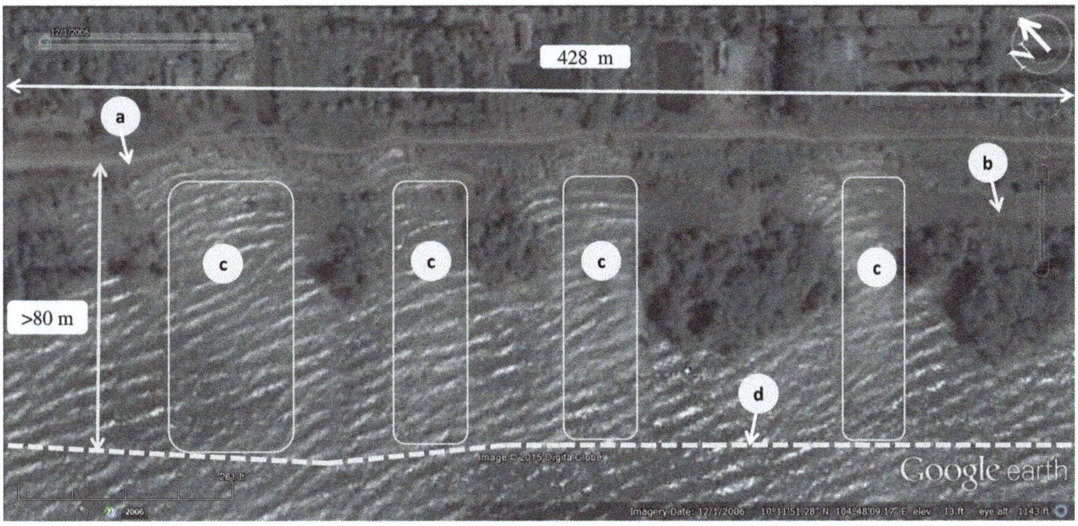

Fig. 2.11 An eroding area with channels dug for local boating in Vam Ray, Hon Dat, Kien Giang. (**a**) breached earthen sea dyke, (**b**) deep channel as a result of excavation of mud and soil for constructing earthen sea dyke, (**c**) channels dug for facilitating local boating, (**d**) mangrove edge in the 1980s. (Adapted from Nguyen et al. (2017))

Fig. 2.12 An eroding earthen sea dyke section between June and July 2009 in Hon Dat, Kien Giang province. (**a**) agriculture production area behind the earthen sea dyke, (**b**) eroding earthen sea dyke section, (**c**) temporary *Melaleuca* fence to protect eroding earthen sea dyke from further erosion, (**d**) eroding area with loss of mangrove trees, (**e**) fragmented mangrove patches

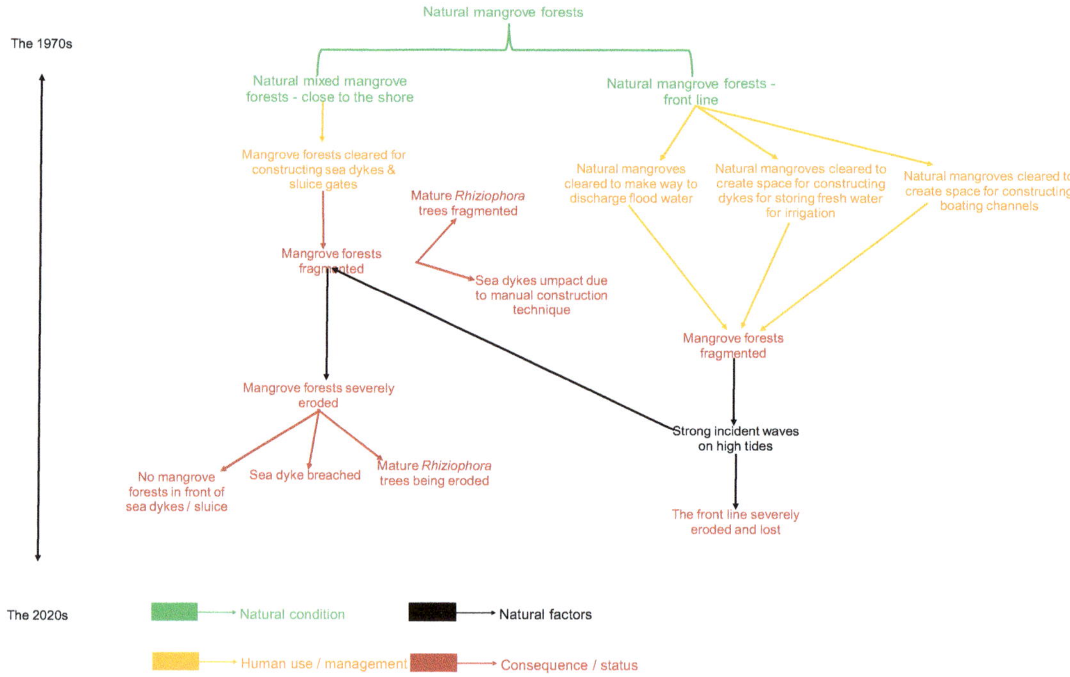

Fig. 2.13 Summary of the erosion process with factors significantly contributing to the erosion process in coastal mangrove protection forests in Kien Giang, VMD. (Adapted from Nguyen (2024))

References

Anthony EJ, Brunier G, Besset M, Goichot M, Dussouillez P, Lap NV (2015) Linking rapid erosion of the Mekong River delta to human activities. Sci Rep 5:14745. https://doi.org/10.1038/srep14745

Besset M, Goichot M, Anthony EJ, Bouchette F, Goichot M, Marchesiello P (2019) Mangroves and shoreline erosion in the Mekong River delta, Viet Nam. Estuar Coast Shelf Sci 226:106263. https://doi.org/10.1016/j.ecss.2019.106263

Bui TL, Hanh HTP (2022) Coastal erosion trend analysis using a combination of remote sensing and hydrodynamic models: case study of Ca Mau Cape, Mekong Delta. Remote Sens Appl 26:100734. https://doi.org/10.1016/j.rsase.2022.100734

Ca Mau PPC (2024) Progress report to the Vietnamese Central Office Working Group on surveying dikes and checking the landslide situation in Ca Mau. Ca Mau PPC, Ca Mau

DARD Ca Mau (2021a) Report on preliminary summary of the implementation of forest allocation in Ca Mau. DARD Ca Mau, Cau Mau

DARD Ca Mau (2021b) Review of forest and forest lands in Ca Mau. DARD Ca Mau, Cau Mau

DARD Ca Mau (2024) Statistics of coastal and riverbank erosion in Ca Mau. DARD Ca Mau, Ca Mau

DARD Kien Giang (2011) Project of coastal mangrove protection forest project and afforestation between 2012 and 2016 in Kien Giang, Vietnam. DARD Kien Giang, Rach Gia

DARD Tien Giang (2020) Report on evaluating climate change and weather forecast in Tien Giang. DARD, My Tho, Tien Giang

DARD Tien Giang (2021) Report on coastal mangrove protection forest. DARD, My Tho, Tien Giang

DARD Tien Giang (2024) Report on coastal mangrove protection forest. DARD, My Tho, Tien Giang

Hanh, Bui (2023) Mechanism of erosion zone formation based on hydrodynamic factor analysis in the Mekong Delta coast, Vietnam. Environ Technol Innov 30:103094. https://doi.org/10.1016/j.eti.2023.103094

Hoa NH, McAlphine C, Pullar D, Leisz SJ, Galina G (2015) Drivers of coastal shoreline change: case study of Hon Dat coast, Kien Giang, Vietnam. Environ Manag 55:1093–1108. https://doi.org/10.1007/s00267-015-0455-7

Kien Giang PPC (2011) Report on the implementation of five year of Decision 51 by Kien Giang PPC on afforestation and forest protection in Kien Giang. Kien Giang PPC, Rach Gia, Kien Giang

Marchesiello P, Minh NT, Gratiot N, Loisel H, Anthony EJ, San DC, Almar R, Kestenare E (2019) Erosion of the coastal Mekong delta: assessing natural against man induced processes. Cont Shelf Res 181:72–89. https://doi.org/10.1016/j.csr.2019.05.004

MARD (2017) Coastal mangrove forests in the Mekong Delta decreased in area over the past 15 years. Available online at: http://vea.gov.vn/vn/quanlymt/baotondadangsh/tintuchoatdong/tintucvasukien/tintuc/Pages/Di%E1%BB%87n-t%C3%ADch-r%E1%BB%ABng-ng%E1%BA%ADp-m%E1%BA%B7n-%C4%90%E1%BB%93ng-b%E1%BA%B1ng-s%C3%B4ng-C%E1%BB%ADu-Long-gi%E1%BA%A3m-trong-5-n%C4%83m-qua.aspx. Accessed 12 Dec 2018

Neuman WL (2011) Social research methods: qualitative and quantitative approaches, 7th edn. University of Wisconsin, Boston

Nguyen TP (2015) Towards sustainable coastal mangrove protection and management in Ben Tre, Vietnam. IUCN, Gland

Nguyen TP (2024) Erosion of mangrove dominated muddy coasts in Vietnam and Indonesia: processes and accelerating factors. Reg Stud Mar Sci 79:103821. https://doi.org/10.1016/j.rsma.2024.103821

Nguyen TP, Parnell KE (2019) Coastal land use planning in Ben Tre: constraints and recommendations. Heliyon 5(4):1–21. https://doi.org/10.1016/j.heliyon.2019.e01487

Nguyen TP, Cottrell A, Parnell KE (2017) Human activities and coastal erosion on the Kien Giang coast, Vietnam. J Coast Conserv 21(6):967–979. https://doi.org/10.1007/s11852-017-0566-9

Nguyen TP, Thuan NB, Tien NN (2025) Monitoring the effectiveness of emergent detached offshore structures in mangrove vegetation increase: lessons and recommendations. Life 15:136. https://doi.org/10.3390/life15020136

Oanh TTK, Lap NV, Tateishi M, Kobayashi I, Saito Y, Nakamura T (2002) Sediment facies and Late Holocene progradation of the Mekong River Delta in Bentre Province, southern Vietnam: an example of evolution from a tide-dominated to a tide- and wave-dominated delta. Sediment Geol 152(3–4):313–325. https://doi.org/10.1016/S0037-0738(02)00098-2

Oanh TTK, Lap NV, Saito Y, Gugliotta M, Tamura T, Lan NTM, Hoang TM, Luan BT (2021) Latest Pleistocene to Holocene stratigraphic record and evolution of the Paleo-Mekong incised valley, Vietnam. Mar Geol 433:106406. https://doi.org/10.1016/j.margeo.2020.106406

Phan LK, Jaap SM, de Vries VT, Stive MJF (2015) Coastal mangrove squeeze in the Mekong Delta. J Coast Res 31(2):233–243. https://doi.org/10.1142/9789814689977_0087

Tien Giang PPC (2017) Report on master plan for development of agro-forestry and aquaculture in Tien Giang until 2020, with vision up to 2030. Tien Giang PPC, My Tho, Tien Giang

Tien Giang PPC (2021) Report on coastal mangrove protection forest. Tien Giang PPC, My Tho, Tien Giang

Tien Giang PPC (2024) Report on coastal mangrove protection forest. Tien Giang PPC, My Tho, Tien Giang

Tra Vinh PPC (2021) Report on preliminary summary of the implementation of forest allocation in Tra Vinh. Tra Vinh PPC, Tra Vinh

Vietnam MOnRE (2020) Scenarios of climate change and sea level rise for Vietnam. Vietnam MOnRE, Hanoi

3 The Legislative Documents That Aim to Sustainably Manage and Use the VMD Coast

Abstract

This chapter provides an overview of the twelve legislative documents intended to promote the sustainable management and use of the VMD coast. Three of these documents are particularly relevant and considered most important when planning the sustainable management and use of the VMD coast for socio-economic development and better adaptation to negative impact of climate change and rising sea levels: Decisions 667, 957 and 17. Decisions 667 and 957 address coastal erosion control, while Decision 17 relates to the protection of coastal mangrove forests. The two coastal erosion control decisions set out the engineering and ecological measures employed to sustainably manage the VMD coast. Engineering solutions include realigning the existing sea dyke system and identifying new locations for constructing sea dykes. These solutions require a focus on utilizing natural features, such as elevated areas and stable geological formations, for reinforcement. They also require consideration of the resettlement and relocation of communities at risk of erosion and the creation of intertidal mudflats for future mangrove planting through the construction of shoreline and offshore measures. Meanwhile, ecological engineering solutions include strict protection of existing sand dunes and intertidal mudflats, as well as strict management of the exploitation of coastal resources. The mangrove protection decision involves sustainable management and use of coastal mangrove protection forests, prevention of forest fires, new forest development, and allocation of mangrove forests to local communities for protection and local improvement through a co-management scheme. Both mangrove protection and new forest development require an understanding of natural conditions and how to apply them during implementation.

Keywords

Coastal mangroves · Coastal protection · Land use · Mangrove · Afforestation sustainable management

3.1 Introduction

Over the past few decades, GoV and the governments at all levels have introduced multiple legal documents that are aimed at sustainably managing VMD and its coast. Sustainable management of VMD and its coast is interpreted as promoting socioeconomic development while effectively controlling the negative impact of climate change, rising sea levels, and coastal erosion.

P. T. Nguyen, *Nature-based Solutions in Mangrove-mud Coast Protection*, Sustainable Development Goals Series, https://doi.org/10.1007/978-981-95-9941-7_3

There were previously four levels of government in Vietnam: the central, the provincial, the city/district, and the communal. In July 2025, the city / district level was removed, leaving Vietnam with three government levels: central, provincial, and communal levels. Cities that are centrally governed cities, such as Ho Chi Minh City, Hanoi City, and Can Tho City, now have the same legislative status as provinces.

According to the Vietnamese legislative system, laws passed by VNA and approved by the President of Vietnam, as well as decisions and decrees issued by VPM, are considered legislative documents. These legislative documents provide administrative and technical guidelines which are further detailed in provincial-level and communal-level decisions and decrees. This chapter therefore focused on these legislative documents to identify legislative documents that are particularly relevant and considered most important when planning the sustainable management and use of the VMD coast. The identification was based on a critical overview of the legislative documents. The identified legislative documents and their applications were then verified through technical discussions with DARD technical staff, field visits and detailed observations.

The Vietnam Portal was selected as the starting point for the critical review because it is operated and maintained by the Vietnam Government Office and contains all the legislative documents that have been issued to date. The aim was to search for policies and legislative documents that regulate the sustainable management, use and protection of the VMD coast.

On *the Vietnam Portal*, the term *Đồng bằng Sông Cửu Long (English version: Vietnamese Mekong Delta—VMD)* was used for retrieving legislative documents in relation to the sustainable management and use of VMD. The search revealed 459 hits. All the hits were classified, analyzed, and limited to the legislative documents and two ecological mangrove management categories: intertidal mudflats and eroding areas and the coastal mangrove protection forests. These two categories reflect natural characteristics of the VMD coast. This classification resulted into 12 hits, which were 8 hits dealing with management and protection of intertidal mudflats and eroding areas and 4 hits relating the management of the two mangrove belts (see Table 3.1).

During the field visits between May 2015 and October 2025, the results from the literature review were verified through the technical discussions with DARD's technical staff of each province (including Section of Irrigation, management boards of coastal mangrove protection forests, and Forestry Section). These discussions were organized at the DARD offices. Additionally, the sites of the existing coastal protection measures as well as their technical designs and their implementation plans were identified and documented to verify the application of the legislative documents in the field and to inform Chap. 4.

Of the twelve (12) legislative documents aimed at managing and protecting intertidal mudflats and eroding areas, three legislative documents: (1) the *Program to Upgrade and Reinforce the Sea Dyke System from Quang Ngai to Kien Giang Provinces* (VPM 2009) (hereafter referred to as Decision 667) and (2) the *Project to Prevent and Control River Bank and Coastal Erosion until 2030* (VPM 2020) (hereafter referred to as Decision 957) and the *Regulations on the management of protection forests* (VPM 2015) (hereafter referred to as Decision 17) are particularly relevant and the most frequently considered during the designing and construction of coastal protection measures and sustainable management and use of the VMD coast for socio-economic development, while the remaining legislative documents provide long term direction and / or vision for sustainable management and use of the VMD as well as the reinforcement of the existing management structures. The first two documents focus on developing solutions to protect intertidal mudflats and prevent control erosion, while the last one provides technical guidelines on classifying and managing protection forests and local involvement in mangrove protection as well as the use of protection forests to improve local livelihoods. Therefore, in addition to the summary of the contents of the remaining documents provided in Table 3.1, this chapter details, analyzes, and summarizes the contents of these three

Table 3.1 Summary of the policies issued by GoV as of October 2025

No.	Policy	Year of issue	Type of legislative document	Overall aim/content
Management and protection of intertidal mudflats and eroding areas				
1	*Strategy for Sustainable Development of Vietnam's Marine Economy to 2030, with a Vision up to 2045* (The Central Committee 2018)	2018	Resolution by the Central Committee	Marine economy to become a strong marine based nation/sustainable marine economy development Better adaptation to climate change and sea level rise Prevention of marine based pollution, environment degradation, coastal erosion and saline intrusion Restoration of, and protection of, important marine ecosystems
2	*Program to Upgrading and Reinforcing the Sea dyke System from Quang Ngai to Kien Giang Provinces* (here in called Decision 667) (VPM 2009)	2009	Decision by VPM	Establishment of a complete and closed sea dike system from Quang Ngai to Kien Giang
3	*Law on natural resources and environment of sea and islands of Vietnam* (VNA 2015)	2015	Law	Integrated management of natural resources of sea and islands based on ecosystem approach and smooth multi-sector coordination
4	*The project on prevention and control of river bank and coastal erosion until 2030* (herein called Decision 957) (VPM 2020)	2020	Decision by VPM	Strengthened management and reinforcement of sea dykes in controlling river bank and coastal erosion
5	*Master plan for fisheries development until 2030, with a vision up to 2050* (VPM 2021a)	2021	Decision by VPM	Development of fisheries to be a key national economy sector. Exploitation of fisheries resources Aquaculture development Processing and exporting fisheries products Coastal development along economic--ecological regions
6	*Program for resettlement of populations in disaster-prone areas, areas with special difficulties, border regions, islands, areas of spontaneous migration, and special-use forests, from 2021 to 2025, with a vision to 2030* (VPM 2022)	2022	Decision by VPM	The program aims to resettle and stabilize residents in areas affected by natural disasters such as erosion, land subsidence, tornadoes, floods, flash floods, inundation, and storm surges The program deals with households and individuals who have lost their homes and land due to landslides, land subsidence, tornadoes, floods, and flash floods, as well as those living in areas at risk of these disasters Resettlement of the population must comply with the planning laws on land, construction, and related plans at all levels Resettlement of the population is a goal and solution for economic, cultural, and social development, as well as for protecting sovereignty, security, national defense, the ecological environment, and water resources

(continued)

Table 3.1 (continued)

No.	Policy	Year of issue	Type of legislative document	Overall aim/content
7	*Strategy for sustainable exploitation and use of natural resources and protection of the environments of seas and islands until 2030 with a Vision up to 2050* (VPM 2023)	2023	Resolution by VPM	Promotion of sustainable use and exploitation of natural resources of seas and islands Maintenance of marine ecosystems and their biological productivity
8	*Laws on fisheries of Vietnam—revised in 2024* (VNA 2024)	2024	Law	Regulations on fishery in land, islands, archipelago and sea of Vietnam and commercial fishing activities outside the Vietnam's maritime boundary
The management and protection of coastal mangrove protection forests				
9	*Strategy for managing the System of Special Use Forests, Marine Protected Areas, Inland Wetland Protected Areas in Vietnam until 2020 with a vision up to 2030* (VPM 2014)	2014	Decision by VPM	Increase of the total area of the systems of special use forests, marine protected areas, inland wetland protected areas Conservation and sustainable development of important natural ecosystems, valuable natural landscapes, endangered, threatened wildlife species in special use forests
10	*Regulations on the management of protection forests* (VPM 2015)	2015	Decision by VPM	Provision of technical guidelines on classifying and managing protection forests
11	*Law on Forestry of Vietnam* (VNA 2017)	2017	Law	Regulation of management, protection, development, and utilization of forests; and the processing and trade of forest products
12	*Strategy for Forestry Development of Vietnam for the period between 2021 and 2030, with a vision up to 2050* (VPM 2021b)	2021	Decision by VPM	Development of forestry into a truly economic and technical sector Establishment, management, protection, development, and sustainable utilization of forests and land areas planned for forestry

legislative documents. The summaries and analyses provide the policy context for the planned sustainable management and use of the VMD coast for socio-economic development.

3.2 The Legislative Documents Aimed to Sustainably Manage and Use the VMD Coast

3.2.1 The Legislative Documents Aimed to Control the Coastal Erosion

Decisions 667 and 957 are two legislative documents that outline engineering and ecological engineering measures used for sustainable management of the VMD coast. Decision 667 plays a crucial role in directing coastal protection and developing erosion control measures in Vietnam. It aims to complete the sea dyke system (including sea dykes and sluice gates) along the coast from Quang Ngai in central Vietnam to Kien Giang in VMD, in an attempt to mitigate the negative impacts of natural factors such as rising sea levels and climate change. This policy also seeks to protect public and private property from erosion, guarantee national defence and security, and create infrastructure that aligns with provincial-level socio-economic development plans. The policy involves two main coastal protection measures: ecological engineering and engineering solutions. Engineering solutions

Fig. 3.1 Summary of the engineering and ecological engineering solutions stipulated by Decision 667

involve realigning the current sea dyke system and identifying new sea dyke routes. These emphasize the use of natural characteristics, such as elevated areas and areas with relatively stable geological foundations, to complete the sea dyke systems. They also emphasize the use of existing structures and mudflats in front of the sea dykes for planting mangrove seedlings, full consideration of the resettlement and relocation of communities prone to erosion, and plans for creating intertidal mudflats for planting trees in front of the sea dykes in the future. Ecological engineering solutions are limited to field investigations, further studies, the strict protection of existing sand dunes and intertidal mudflats, the strict management of coastal resource exploitation, and monitoring and evaluation. This decision is summarized in Fig. 3.1.

Decision 957 aims to provide full control over coastal and riverbank erosion in the VMD and Central Coastal Vietnam up to 2030. The policy aims to proactively manage and control coastal and riverbank erosion in these areas. This is achieved through critical reviews and finalization of legal documents on coastal and riverbank erosion, and by undertaking basic scientific studies on erosion to develop a comprehensive database and erosion control plans, as well as implementing solutions to control erosion. Urgent solutions include erection of warning signs in erosion hotspots, strict management of coastal and riverine resource exploitation, resettlement and relocation of communities at risk of erosion, construction of erosion control structures, strict control of construction and renovation of houses along riverbanks and coasts, and effective management of waterborne transport vehicles. Long-term solutions include interpretation, education and awareness, master plans for the resettlement of coastal and riverbank communities, proactive relocation of communities at risk of erosion, development of erosion process warning systems, construction of erosion control structures at hotspots to protect public property and lives, implementation of coastal mangrove afforestation projects, and the use of science and technology for developing erosion control measures. The decision is summarized in Fig. 3.2.

3.2.2 The Legislative Document to Protect and Manage the Coastal Mangrove Protection Forests

Decision 17 stipulates the regulations of management of protection forests in Vietnam. The protection forests in Vietnam are classified into four categories, including protection forests for watershed protection, protection forests for windbreak

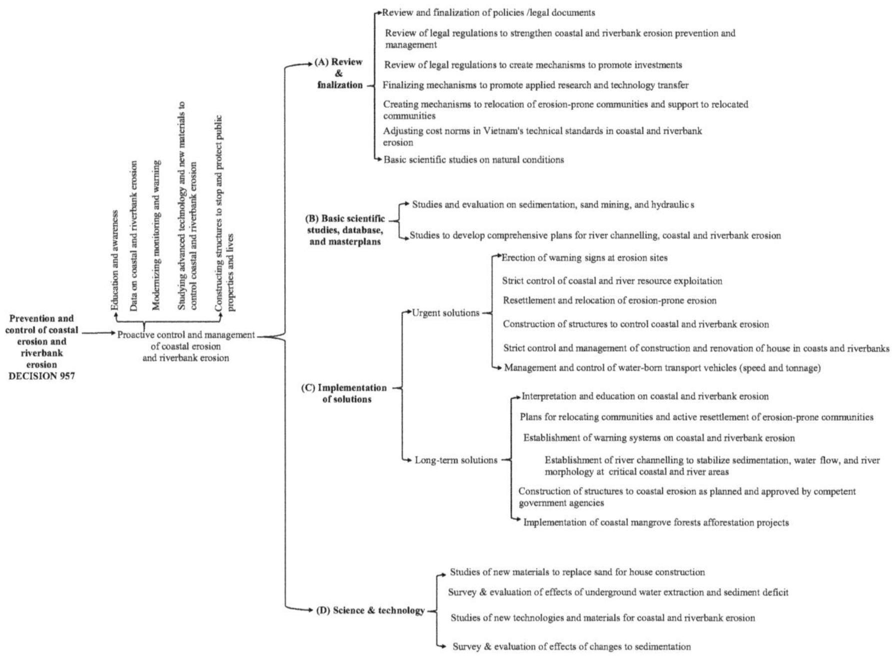

Fig. 3.2 Summary of the engineering and ecological engineering solutions stipulated by Decision 957

and sand fixing, protection forests for wave dissipation and land reclamation, and protection forests for environmental protection. Each protection forest is legally managed by a management board, which is established by PPC of each province. Coastal mangrove forests are classified as protection forests for the purposes of wave dissipation and land reclamation (hereafter referred to as coastal mangrove protection forests).

Decision 17 applies to coastal mangrove protection forests in VMD (please see Chap. 1 for further information regarding the management of coastal mangrove protection forests). Under this decision, management boards of these forests are responsible for protecting the existing mangrove forests and allocating coastal mangrove protection forests to local communities for co-management and livelihood improvement. The protection of the existing mangrove forests involves protection of mangrove species, prevention of forest fires, and development of new mangrove forests.

Co-management is undertaken through 49-year contracts. In these contracts, the management boards of coastal mangrove protection forests (called the contractors) sign with local communities that depend on mangrove resources for their livelihood, as well as with government agencies such as the army, public security agencies, the Women's Union, and the Farmers' Union (called the contractees). The coastal mangrove protection forests, which are contractually allocated, are located in the secondary mangrove belt. Contractors are responsible for preparing administrative procedures, including the preparation of contracts and the development of guidelines for the use of allocated forests and afforestation. They are also responsible for monitoring the allocation processes and undertaking annual performance and evaluation. They are solely responsible for protecting and managing unallocated mangrove forests, which are mainly located in the primary mangrove belt (Nguyen and Nuong 2023).

In return for the use of the forests for aquaculture development and livelihood improvements, contractees are legally responsible for protecting up to 70% of the allocated forests. Aquaculture development can be undertaken in the remaining 30%. This includes farming mud crabs, shrimp,

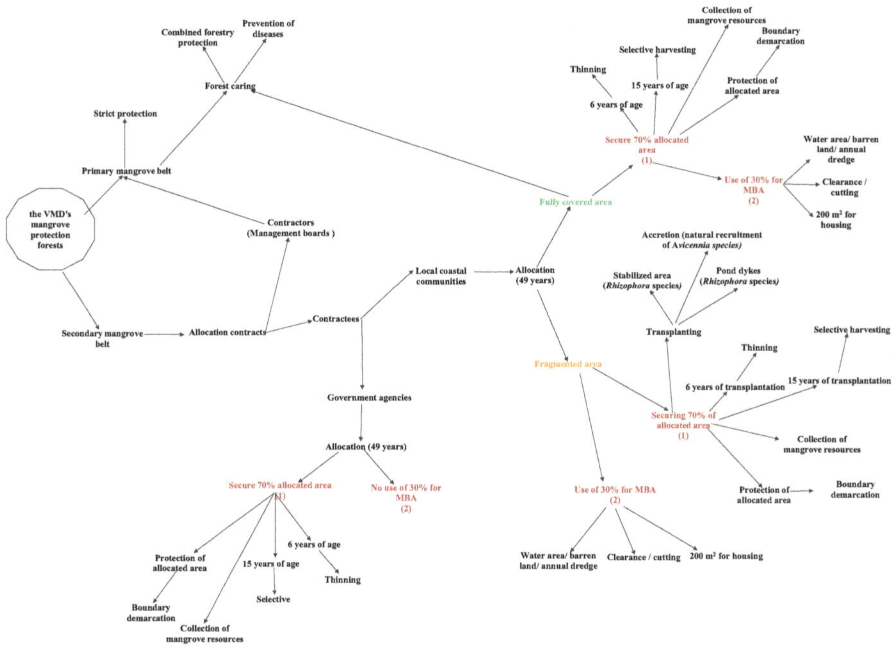

Fig. 3.3 Summary of the allocation of coastal mangrove protection forests for protection and livelihood improvement. (Adapted from Nguyen and Nuong 2023; Nguyen et al. 2023). For details on the allocation program, please refer to Nguyen and Nuong (2023); Nguyen et al. (2023)

barramundi, mollusks, and other marine fish species. However, contractees are responsible for planting mangrove seedlings to create new mangrove forests. If the allocated mangrove forests are barren or fragmented, the contractees must plant mangrove seedlings in these areas annually to achieve the required mangrove coverage of 70%. Contractees can also plant seedlings on pond dikes, which count toward the required mangrove coverage (Nguyen and Nuong 2023).

In addition to products of aquaculture development, contractees are entitled to products from the thinning of the allocated mangrove forests after 6 years of planting and from selective harvesting of these forests after 15 years of planting. They are also entitled to collect mangrove resources, including fish, crabs, and mollusks. Contractees are also paid to protect 70% of the allocated mangrove forests, plant seedlings, and replace dead ones. Contractees being government agencies have been involved in protecting these mangrove forests and planting seedlings to create new ones. However, they are not heavily involved in aquaculture development. In contrast, contractees who are households are heavily involved in aquaculture development (Nguyen and Nuong 2023).

In 2024, many VMD provinces stopped signing new allocation contracts as well as prohibited clearance of allocated mangrove forests for aquaculture development in accordance with Vietnam's Forestry Law. Other contractual requirements such as protection of existing mangrove forests and development of new mangrove forests remained unchanged (VNA 2017) (Fig. 3.3).

3.3 Conclusions

This chapter provides a critical review and summary of the twelve legislative documents aimed at the sustainable management of the VMD coast. Three of these documents are particularly relevant, as they are closely related to the sustainable management and effective use of coastal mangrove protection forests on the VMD coast. These documents are considered the most important when planning the sustainable management and use of the VMD coast for socio-economic development. Decisions 667 and 957 are crucial for effectively protecting the VMD coast, managing coastal mangrove protection forests, and controlling coastal and riverbank erosion. Both decisions emphasize the importance of taking natural conditions into account when planning and constructing coastal structures. This is achieved through basic scientific studies, finalizing legislative documents and policies, and encouraging local involvement in mangrove protection activities. The resettlement programme is an essential element of sustainable coastal protection, helping to stabilize economic, cultural and social development, as well as protecting sovereignty, security, national defence, the ecological environment and water resources. Decision 17 stipulates the sustainable management and use of coastal mangrove protection forests for socio-economic development, as well as for adapting to the negative impact of climate change and rising sea levels. This is ensured through forest protection by management boards, the allocation of coastal mangrove protection forests for co-management and the development of new forests and improved livelihoods.

References

Nguyen TP, Nuong TC (2023) Thinning and selective harvesting and mangrove protection forests: lessons learned and recommendations from the Vietnamese Mekong Delta. Estuar Coast Shelf Sci 288:108345. https://doi.org/10.1016/j.ecss.2023.108345

Nguyen TP, Nuong TC, Quang NH (2023) Local perceptions of mangrove protection and livelihood improvement in co-management: lessons learnt and recommendations. Ocean Coast Manag 237:106530. https://doi.org/10.1016/j.ocecoaman.2023.106530

The Central Committee (2018) Strategy for sustainable development of Vietnam's marine economy to 2030, with a vision up to 2045. The Central Committee, Hanoi

VNA (2015) Law on natural resources and environment of sea and islands of Vietnam. VNA, Hanoi

VNA (2017) Law on Forestry of Vietnam. VNA, Hanoi

VNA (2024) Laws on fisheries of Vietnam – revised in 2024. VNA, Hanoi

VPM (2009) Decision on approving reinforcing and upgrading sea dyke systems between Quang Ngai and Kien Giang. VPM, Hanoi

VPM (2014) Strategy for managing the system of special use forests, marine protected areas, inland wetland protected areas in Vietnam until 2020 with a vision up to 2030. VNA, Hanoi

VPM (2015) Regulations on the management of protection forests. VNA, Hanoi

VPM (2020) The project on prevention and control of river bank and coastal erosion until 2030. VPM, Hanoi

VPM (2021a) Master plan for fisheries development until 2030, with a vision up to 2050. VPM, Hanoi

VPM (2021b) Strategy for Forestry Development of Vietnam for the period between 2021 and 2030, with a vision up to 2050. VNA, Hanoi

VPM (2022) Program for resettlement of populations in disaster-prone areas, areas with special difficulties, border regions, islands, areas of spontaneous migration, and special-use forests, from 2021 to 2025, with a vision to 2030. VPM, Hanoi

VPM (2023) Strategy for sustainable exploitation and use of natural resources and protection of the environments of seas and islands until 2030 with a vision up to 2050. VPM, Hanoi

4 The Existing Measures to Protect the VMD Coast

Abstract

This chapter illustrates the existing coastal protection measures, which have been constructed along the VMD coast up to date. These measures included mangrove afforestation and the construction of shoreline and offshore structures. Mangrove afforestation involved planting of seeds/propagules and seedlings that are at 6 months old. Seeds/propagules were planted at a high density of 20,000 seeds or propagules per ha while seedlings were planted at a less density of 2500 seedlings and with support provided by rows of poles, made of bamboo, *Eucalypt* fences, or *Melaleuca* sticks, concrete cement. The construction of shoreline structures included the construction of sea dykes and revetments to protect the VMD coast from further erosion. Offshore structures, such as Geo-tubes, offshore rock revetments, and offshore concrete breakwaters were installed parallel to the shore to dissipate strong wave energy, allowing sediment to accumulate and mangrove species to grow.

Keywords

Mangrove afforestation · Offshore breakwater · Rock revetments · Sea dykes

4.1 Introduction

Between 2009 and 2025, several efforts were made to implement Decisions 667, 957 and 17, with the aim of sustainably managing and using the VMD coast. These efforts included the construction of shoreline measures (sea dykes and sluice gates) and offshore structures (offshore breakwaters) and mangrove afforestation.

This chapter provides an overview of the existing measures implemented along the VMD coast, in order to understand what measures were constructed and how they were developed in terms of design and construction. Lessons will be drawn from this overview for future coastal management and control. As most of the designs and implementations were filed as progress reports, technical reports and technical designs in the DARD offices, the overview analyzed this grey

P. T. Nguyen, *Nature-based Solutions in Mangrove-mud Coast Protection*, Sustainable Development Goals Series, https://doi.org/10.1007/978-981-95-9941-7_4

Table 4.1 Summary of articles/documents retrieved from grey and peer-reviewed literature

No.	Documents (year)/articles (journal) Grey literature	Department/ author(s)
1	Coastal rehabilitation and mangrove restoration using *melaleuca* fences: practical experience from Kien Giang Province	GTZ (2012a)
2	Erosion management and mangrove restoration plan for the mainland coast of Kien Giang, Vietnam in 2012	GTZ (2012b)
3	Shoreline and mangrove resource condition of the mainland coastline of Kien Giang Province, Vietnam: key pressures and restoration requirements in 2012	GTZ (2012c)
4	Effectiveness of *melaleuca* fences for mangrove restoration rehabilitation efforts in 2012	GTZ (2012d)
5	Report on evaluating climate change and weather forecast in Tien Giang	Tien Giang DARD (2020)
6	Coastal protection for the Mekong Delta: a decision support tool	GTZ (2018)
7	Decision on adjusting the investment project to tackle urgently coastal erosion in Mui Ranh, An Bien district, Kien Giang province	Kien Giang PPC (2021a)
8	Decision on the investment project to tackle urgently coastal erosion in the coastal area from Chu Vang to Muoi Than, An Minh district, Kien Giang province	Kien Giang PPC (2021b)
9	Decision on adjusting the investment project to tackle urgently coastal erosion in Hon Queo area, Hon Dat district, Kien Giang province	Kien Giang PPC (2021c)
10	Decision on adjusting the investment project to tackle urgently coastal erosion in Xeo Nhau area, An Minh district, Kien Giang province	Kien Giang PPC (2021d)
11	Project document: integrated coastal protection and mangrove restoration in Kien Giang and Ca Mau provinces	MARD (2022)

literature, and conducted semi-structured interviews with technical staff at the DARD offices, making detailed observations and taking documentation during field visits between May 2019 and October 2025. Semi-structured interviews were conducted using summaries and analyses of Decisions 667, 957 and 17 to stimulate discussion, as presented in Chapter 3. The results of the semi-structured interviews were verified using photographs, detailed observations and field documentation (Table 4.1).

4.2 The Existing Measures to Protect the VMD Coast

The semi-structured interviews revealed that multiple coastal protection measures had been implemented as pilot demonstrations, including *Melaleuca* fences (Nguyen et al. 2016), T-shaped fences in Soc Trang (Schmitt et al. 2013; Schmitt and Albers 2014), and bamboo fences in Bac Lieu (Mai et al. 2021). However, most of these initiatives were no longer visible on the sites during the field visits. This was possible due to high levels of sediment accumulation and the robust growth of mangrove forests, burying these measures completely (Nguyen et al. 2016), or them being washed away (Schmitt and Albers 2014). As of October 2024, existing measures included mangrove afforestation, the construction of shoreline and offshore structures to protect the VMD from further erosion as well as to enable the sustainable management and use of the VMD coast and improve its ability to adapt to the negative impacts of climate change and rising sea levels.

4.2.1 Mangrove Afforestation

The management boards included mangrove afforestation as a crucial element of their annual working plans to develop new mangrove forests. This was often undertaken in strictly protected areas of coastal mangrove protection forests or in areas allocated to contractees for aquaculture development (please see Chap. 3 for additional information regarding the allocation program). Mangrove afforestation involved the planting of propagules/seeds or planting of mangrove seedlings which were nurtured in nurseries, which is regulated by MARD (2016).

Fig. 4.1 Mangrove seedlings were planted in enclosed bay or in river mouths. The left photo shows planting of Avicennia seedlings in an enclosed bay in Ca Mau. The right one shows planting in a river mouth area in Tra Vinh. (**a**) standing *Avicennia* trees, (**b**) planted *Avicennia* 1-year seedlings, (**c**) offshore bamboo fences with steel frame support, (**d**) 3-year *Bruguiera* seedlings planted in a riverine area, (**e**) *Eucalypt* fences erected offshore to protect *Bruguiera* seedlings

Seeds/propagules (usually *Avicennia* or *Rhizophora* species) were often collected in areas where mother trees are growing and then transported to designated planting sites for planting. They were planted at a density of up to 20,000 seeds/propagules per ha in low energy environments and/or intertidal mudflats.

The planting of seedlings involved the production of seedlings in nurseries. The production started with collecting and transporting wild seeds/propagules to nurseries. These nurseries were constructed near designated planting sites to facilitate easy logistics and preparation. The first step in the nursery operation was to fill standard pots (plastic bags) with compost. The second step involved potting the propagules, seeds, or seedlings in the compost pots. The potted seedlings were irrigated daily with freshwater in nurseries for at least 6 months. Popular seedlings were seedlings of *Avicennia*, *Bruigueria*, or *Rhizophora* species.

Seedlings were planted with support provided by bamboo, *Eucalypt* fences, or *Melaleuca* sticks, concrete poles, and at different densities per ha, depending on the mangrove species. Seedlings of *Avicennia Rhizophora* species were planted in prograding areas, intertidal mudflats, or fragmented mangrove areas, or in enclosed bays, to ensure high survival rates. Seedlings of *Bruigueria* species were planted in estuaries (Fig. 4.1).

Planting was usually outsourced due to insufficient staffing of the management boards. Outsourced personnel could be external agencies or individuals with extensive planting experience, or community members who lived in areas near the planting sites were willing to participate in and maintain the planting sites. Planted seedlings were inspected one or two months after planting. If survival rate were less than 10% of the seedlings die, replanting was not be necessary; Otherwise, it was. Replanting was also required during the first three years, with a replanting rate of at least 15% in the first year, 10% in the second year, and 5% in the third year.

In addition to their use in low-energy environments, planting seedlings was previously an important measure for erosion control. However,

Fig. 4.2 Mangrove seedlings with *Malaleuca* sticks died in an eroding area in Hon Dat, Kien Giang, Vietnam. (**a**) *Melaleuca* sticks remained on the spot after seedlings were washed away. (**b**) *Eucalypt* fences were constructed offshore to protect seedlings from erosion. (The photo was taken in May 2009)

this method often failed in eroding areas because the seedlings could not withstand strong waves and were uprooted and washed away (Fig. 4.2).

4.2.2 Construction of Shoreline Structures

The construction of infrastructure for agriculture, forestry, and rural development, including sea dykes, revetments, offshore breakwaters, irrigation systems, and reservoirs, is currently subject to four administrative phases. These are (1) undertaking of surveys and planning, (2) preparation and approval of technical designs, (3) construction, and (4) operation and maintenance. In the first phase, DARD establish management boards for agricultural and rural development projects. These management boards are responsible for conducting surveys and planning. The management boards report survey findings and prepare calls for open bidding. The second phase begins with the selection of bidders who meet all the requirements. Successful bidders then proceed with the technical design and construction plans, which are then submitted to these management boards for approval. The technical design of the sea dykes is undertaken in accordance with the Dykes Law (VNA 2025) and is approved by central government agencies. Once construction is complete and approved, the infrastructure undergoes technical and financial inspections and approvals before being transferred to the management boards of coastal mangrove protection forests for operation and maintenance.

The construction of shoreline structures included construction of sea dykes and revetments along the VMD coast. In reality, these structures were constructed as short-term strategies in areas experiencing active erosion to prevent further erosion. Existing earthen sea dykes were upgraded and reinforced with cement blocks to widen the crest for local transportation purposes. New sea dikes were constructed using sediment excavated from both sides as well as cement blocks of various sizes, depending on local preference (Nguyen et al. 2017). Sluice gates were built across existing channels and small rivers as a crucial part of the sea dyke system in VMD. There were different technical designs of sluice gate. The crest of the sea dyke was +3 m high and was used for local transportation in most provinces (Fig. 4.3).

Revetments were constructed using either granite rocks glued with cement or iron nets containing granite rocks. Alternatively, pile sheets were strengthened with iron poles and wires. In some provinces, revetments were constructed using reinforced cement blocks that

Fig. 4.3 Sea dykes and sluice gates in Hon Dat, An Giang. (**a**) and (**b**) are sluice gates. (**c**) and (**d**) are sea dykes with fabricated cement blocks on a slope facing the western sea

Fig. 4.4 Shoreline revetments in Dong Thap and Ca Mau provinces. (**a**) rock revetment constructed to close completely mangrove areas behind. (**b**) revetments were constructed with Tetrapods in Bac Lieu. (**c**) rock and iron wire net revetments in Ca Mau. (**d**) pile sheets with iron poles reinforced with iron wires in Ca Mau

Fig. 4.5 Revetments in Dong Thap and Ca Mau provinces. (**a**) geotubes constructed about 200 m off of the Go Cong Dong shoreline, Dong Thap. (**b**) rock revetment constructed approximately 150 m off the Go Cong Dong shoreline, Dong Thap. (**c**) and (**d**) are iron wire granite rock revetments in Ca Mau

were protected by reinforced cement tetrapods. On the eastern side of VMD, tetrapod revetments were popular, while on the western side, iron wire-rock revetments were a common coastal protection measure (Fig. 4.4).

4.2.3 Construction of Offshore Breakwaters with or Without Planting of Mangrove Seedlings

Geotubes and concrete offshore breakwaters were constructed along the coasts of the provinces of Dong Thap, Vinh Long, and Ca Mau. The planting of mangrove seedlings (mainly with *Avicennia* species) was originally a component of these measures, but this was not undertaken until sediment had accumulated around. However, there were no signs of sediment accumulation. Consequently, the planting was not implemented (Fig. 4.5).

From 2016 to 2023, different types of offshore breakwater were constructed in certain provinces of VMD to achieve the objectives set out in Decision 957 (please refer to Chap. 1 for further information on offshore structures). These structures were constructed off the shoreline to create space for planting mangrove seedlings at a later stage. These structures were usually designed by research institutes and/or central government agencies. These institutes were responsible for designing the structures and search for funding and testing these structures in provinces. The distance between the offshore structures and shorelines varied from province to province.

Funding was normally provided through scientific research and technology programs at the central and provincial levels. In Vietnam, funds were available through the National level Fund for Science and Technology Development in Vietnam (Nafosted), which was run by Ministry of Science and Technology or funds through science and technology development program run

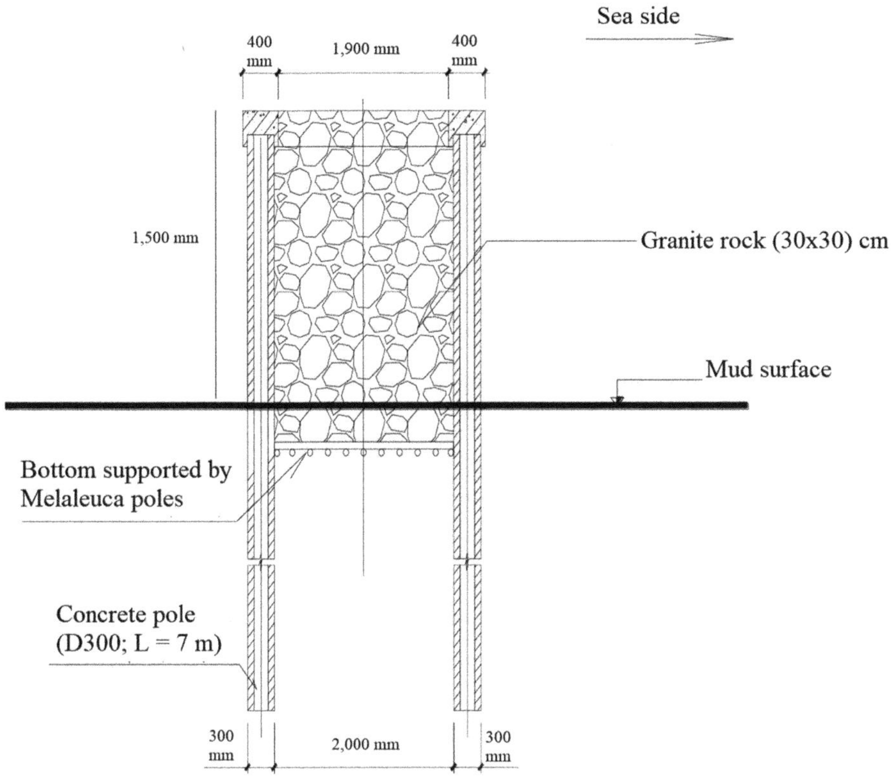

Fig. 4.6 Technical design of double-line granite rock breakwater in Ca Mau province, VMD. (Adapted from Luom et al. (2021) and Nguyen (2022))

by provincial departments of science and technology.

Ca Mau was the first province in Vietnam to construct different types of offshore structures, including pile-rock breakwaters and hollow cylinder breakwaters. Approximately 145 km of pile-rock breakwaters were constructed along the Ca Mau province. Pile-rock breakwaters consisted of two lines of concrete piles spaced 2.0 m apart, with granite rocks filling the gaps. The crest was at the height of about +1.5 m, which is higher than the highest tides on the western side (+1 m). These structures have never been overtopped, not even by the highest tides. *Avicennia* seedlings were planted in the gaps between the pile-rock breakwaters and the shoreline. By 2020, the eroding coast had stabilized thanks to the protection of existing and naturally growing mangrove trees (Luom et al. 2021) (Fig. 4.6).

In 2020, a hollow cylinder breakwater was also tested in Vam Da Bac, Ca Mau province. However, there was limited information about its effectiveness in controlling the erosion in the area (Nguyen 2022). Between 2020 and 2023, hollow triangular breakwaters were constructed along the Tien Giang coast. No seedlings were planted in the gaps between the hollow triangular breakwaters and Tien Giang shoreline (Nguyen et al. 2025) (Figs. 4.7 and 4.8).

4.3 Conclusions

This chapter summarizes the coastal protection measures which were installed along the VMD coast. These measures included mangrove afforestation (mainly *Rhizophora*, *Bruigueria* and *Avicennia* species), the construction of sea dykes and sluice gates, shoreline and offshore revet-

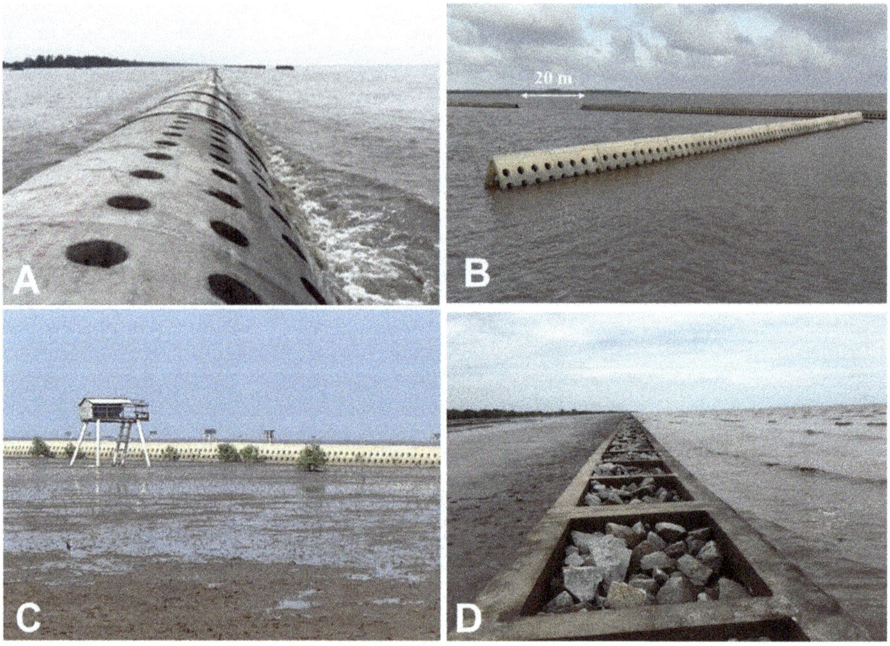

Fig. 4.7 Offshore structures in Tien Giang and Ca Mau provinces. (Adapted from Nguyen (2022)). (**A**) hollow cylinder breakwaters in Vam Da Bac, Ca Mau. (**B**) hollow triangular breakwaters in Con Cong, Tien Giang. (**C**) hollow triangular breakwaters in Tan Thanh, Tien Giang. (**D**) pile rock breakwaters in Huong Mai, Ca Mau. By 2025, there was a sign of naturally growing mangrove trees (mainly *Avicennia* species) in areas in front of and right behind these structures (Nguyen et al. 2025)

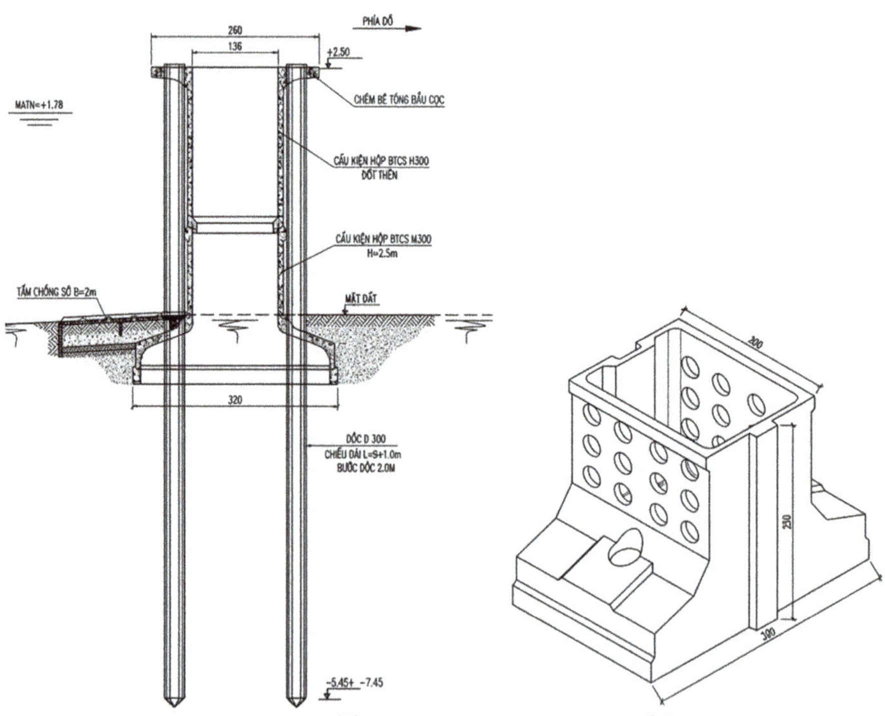

Fig. 4.8 Technical design of offshore structures in Tien Giang (**a**) technical design (**b**) cross section. (Adapted from Nghia et al. 2020; Tu et al. 2020, 2022a, b, c). By 2025, there was a sign of naturally growing mangrove trees (mainly *Avicennia* species) in areas in front of and right behind these structures (Nguyen et al. 2025)

ments, and offshore breakwaters (including pile rock breakwaters, hollow triangular breakwaters, hollow triangular breakwaters). Mangrove afforestation was usually undertaken in low-energy environments such as intertidal mudflats, enclosed bays, fragmented mangrove areas, and in areas protected by additional measures such as *Melaleuca*, bamboo, concrete poles. However, mangrove afforestation failed in eroding areas due to poor seedling quality and inappropriate planting techniques. Shoreline structures included construction of sea dykes through a new construction process as well as the upgrading of current earthen sea dykes to protect the sites from further erosion. The construction of offshore breakwaters was primarily intended to dissipate the energy of strong incident waves, sometimes with planting of seedlings of *Avicennia* and *Rhizophora* species. However, there was limited evidence of mangrove growth behind these offshore structures.

References

GTZ (2012a) Coastal rehabilitation and mangrove restoration using melaleuca fences: practical experience from Kien Giang Province. CDBRP, Rach Gia, Kien Giang

GTZ (2012b) Erosion management and mangrove restoration plan for the mainland coast of Kien Giang, Vietnam in 2012. CDBRP, Rach Gia, Kien Giang

GTZ (2012c) Shoreline and mangrove resource condition of the mainland coastline of Kien Giang Province, Vietnam: key pressures and restoration requirements in 2012. CDBRP, Rach Gia, Kien Giang

GTZ (2012d) Effectiveness of melaleuca fences for mangrove restoration rehabilitation efforts in 2012. CDBRP, Rach Gia, Kien Giang

GTZ (2018) Coastal protection for the Mekong Delta: a decision support tool. GTZ, Hanoi

Kien Giang PPC (2021a) Decision on adjusting the investment project to tackle urgently coastal erosion in Mui Ranh, An Bien district, Kien Giang province. Rach Gia, Kien Giang, Kien Giang PPC

Kien Giang PPC (2021b) Decision on the investment project to tackle urgently coastal erosion in the coastal area from Chu Vang to Muoi Than, An Minh district, Kien Giang Province. Kien Giang PPC, Rach Gia, Kien Giang

Kien Giang PPC (2021c) Decision on adjusting the investment project to tackle urgently coastal erosion in Hon Queo area, Hon Dat district, Kien Giang Province. Kien Giang PPC, Rach Gia, Kien Giang

Kien Giang PPC (2021d) Decision on adjusting the investment project to tackle urgently coastal erosion in Xeo Nhau area, An Minh district, Kien Giang Province. Kien Giang PPC, Rach Gia, Kien Giang

Luom TT, Nguyen TP, Anh NT, Tung NT, Tu LX, Duong TA (2021) Using fine-grained sediment and wave attenuation as a new measure for evaluating the efficacy of offshore breakwaters in stabilizing an eroded muddy coast: insights from Ca Mau, the Mekong Delta of Vietnam. Sustainability 13(9):4798. https://doi.org/10.3390/su13094798

Mai CV, Ngo A, Mai T, Dao HT (2021) Bamboo fences as a nature-based measure for coastal wetland protection in Vietnam. Front Mar Sci 8:756597. https://doi.org/10.3389/fmars.2021.756597

MARD (2016) Technical guidelines on planting six mangrove species: Avicennia marina, Avicennia alba, Rhizophora apiculata, Rhizophora mucronate, Sonneratia alba, and Lumnitzera littorea. MARD, Hanoi

MARD (2022) Project document: integrated coastal protection and mangrove restoration in Kien Giang and Ca Mau provinces. MARD, Hanoi. Available online at https://mic.mard.gov.vn/Project/3047.aspx. Accessed 12 Dec 2024

Nghia NV, Minh HVT, Luan TC, Ty TV (2020) Assessment of wave reduction through the Busadco breakwaters: a case study in the East and West coasts of Ca Mau. Viet J Const 6:584–591

Nguyen TP (2022) Offshore structures, sea mud accumulation and mangrove regeneration: insights from the Vietnamese Mekong Delta, Vietnam. Ocean Coast Manag 225:106243. https://doi.org/10.1016/j.ocecoaman.2022.106243

Nguyen TP, Tam NV, Quoi LP, Parnell KE (2016) Community perspectives on an internationally funded mangrove restoration project: Kien Giang Province, Vietnam. Ocean Coast Manag 119:146–154. https://doi.org/10.1016/j.ocecoaman.2015.10.008

Nguyen TP, Cottrell A, Parnell KE (2017) Human activities and coastal erosion on the Kien Giang coast, Vietnam. J Coast Conserv 21(6):967–979. https://doi.org/10.1007/s11852-017-0566-9

Nguyen TP, Thuan NB, Tien NN (2025) Is it ecologically feasible to enclose mangrove-mud coasts using multiple concrete structures for erosion control? Evidence from the Vietnamese Mekong Delta. Ecol Eng 219:107678. https://doi.org/10.1016/j.ecoleng.2025.107678

Schmitt K, Albers T (2014) Area coastal protection and the use of bamboo breakwaters in the Mekong Delta. In: Coastal disasters and climate change in Vietnam: engineering and planning perspectives. Elsevier, pp 107–132. https://doi.org/10.1016/B978-0-12-800007-6.00005-8

Schmitt K, Albers T, San DC (2013) Site-specific and integrated adaptation to climate change in the coastal mangrove zone of Soc Trang Province, Viet Nam. J

Coast Conserv 17:545–558. https://doi.org/10.1007/s11852-013-0253-4
Tien Giang DARD (2020) Report on evaluating climate change and weather forecast in Tien Giang. Tien Giang DARD, My Tho
Tu LX, Ba HT, Le Manh H, Van DD, Nguyen NM, Wright DP, Bui VH, Mai ST, Anh DT (2020) Hydraulic performance and wave transmission through pile-rock breakwaters. Ocean Eng 218:108229. https://doi.org/10.1016/j.oceaneng.2020.108229
Tu LX, Ba TH, Thanh VQ, Wright DP, Tanim AH, Anh DT (2022a) Evaluation of coastal protection strategies and proposing multiple lines of defense under climate change in the Mekong Delta for sustainable shoreline protection. Ocean Coast Manag 228:106301. https://doi.org/10.1016/j.ocecoaman.2022.106301
Tu LX, Manh LH, Ba HT, Duong DV, Vu HTD, Wright D, Bui VH, Anh DT (2022b) Wave energy dissipation through a hollow triangle breakwater on the coastal Mekong Delta. Ocean Eng 245:110419. https://doi.org/10.1016/j.oceaneng.2021.110419
Tu LX, Phong NC, Thanh VQ, Quan TQ, Wright DP, Anh DT (2022c) Multi-scale modelling for hydrodynamic and morphological changes of breakwater in coastal Mekong Delta in Vietnam. J Coast Conserv 26:18. https://doi.org/10.1007/s11852-022-00866-3
VNA (2025) The law on sea dykes. VNA, Hanoi

5 Natural Processes Along the VMD Coast

Abstract

This chapter illustrates the natural processes that have taken place in coastal environments along the VMD coast. These processes included the morphological evolution of intertidal mudflats and the natural growth of mangrove species in areas experiencing severe erosion. The evolution of intertidal mudflats was recorded in two natural coastal environments, and the growth of the mangrove species was observed in two areas experiencing severe erosion along the VMD coast. The evolution of these intertidal mudflats depended on favourable hydrological characteristics, such as low incident waves, a sufficient supply of fine-grained sediment, and the growth of mangrove trees from floating seeds. The natural mangrove growth in eroding areas was influenced by floating mangrove seeds and microsites that entrapped marine debris and the aerial roots of mangrove species. These processes were not complete until human exploitation of coastal resources was eliminated.

Keywords

Entrapping sites · Intertidal mudflats' · Mangrove propagules · Natural growth

5.1 Introduction

Mangrove-mud coasts evolve through the accumulation of sufficient sediments and the robust growth of mangrove species. Sediment accumulation and mangrove growth are widely reported in the literature as positive results of coastal protection solutions, including ecological engineering solutions and engineering solutions. However, little is known about the processes by which sediment accumulates and mangrove species grow in natural environments, particularly in areas where coasts are eroding. This knowledge is crucial, particularly given the increased recommendations of NbS as a cost-effective way to control the erosion of mangrove muddy coasts (Perricone et al. 2023; Wu et al. 2023; Yasmeen et al. 2024). Since NbS focuses on the services that nature provides in order to establish effective interventions to improve the quality of modified ecosystems (UNEP 2020), understanding these natural processes provides a comprehensive basis for solutions that utilize the services that nature provides to control the erosion of mangrove-mud coasts.

Despite severe erosion, natural processes have taken place over the VMD coast over the past few years. The processes included the natural growth of mangroves and the evolution of intertidal mud-

P. T. Nguyen, *Nature-based Solutions in Mangrove-mud Coast Protection*, Sustainable Development Goals Series, https://doi.org/10.1007/978-981-95-9941-7_5

flats. Meanwhile, the existing coastal erosion control measures have only had limited success in protecting the coast from further erosion. This limited success was due to a lack of knowledge regarding the morphological characteristics of mangrove-mud coasts (Nguyen et al. 2025b). This chapter therefore documents the evolution of intertidal mudflats and the natural growth of mangrove species along the VMD coast. This documentation is based on a critical review of peer-reviewed publications and field surveys conducted by the author along the VMD coast between 2009 and 2025, the photographs taken during the field visits as well as the author's own publications.

The first step was to critically review papers published in three major domains, such as *SpringerLink*, *ScienceDirect*, and *MDPI* data sources that were critically reviewed to understand the establishment of intertidal mudflats and natural growth of mangrove species along the VMD coast. The search terms used in the review were "*intertidal mudflats in the Vietnamese Mekong Delta*," "*natural growth of mangrove species in the Vietnamese Mekong Delta*." The author recognized that this selection of these search terms could have missed relevant hits, but it helped the author retrieve the correct articles in pursuit of the objective of the study. All hits were classified and analyzed using thematic analysis and to avoid duplication (Table 5.1).

All hits were analyzed and synthesized using participatory diagramming method recommended by Neuman (2011). The analysis and synthesis resulted in a comprehensive insight into natural processes and sites along the VMD. In addition, Google Earth software was used for

Table 5.1 Summary of articles retrieved from a critical review of the main domains

Source	Search term	Hits
ScienceDirect	*Intertidal mudflats in the Vietnamese Mekong Delta*	41
SpringerLink		9
MDPI data source		0
ScienceDirect	*Natural growth of mangrove species in the Vietnamese Mekong Delta*	146
SpringerLink		62
MDPI data source		0

Table 5.2 Summary of articles retrieved from a critical review of the main domains

No.	Paper/sources	Content	Remarks
	Evolution of intertidal mudflats		
1	Protected Mangrove Forests and Aquaculture Development for Livelihoods (Luom et al. 2021)	Intertidal mudflats in low-energy environment in An Bien, An Giang, VMD	The western side of VMD
2	Thinning, Selective Harvesting, and Mangrove Protection Forests: Lessons Learned and Recommendations from the Vietnamese Mekong Delta (Nguyen and Nuong 2023)	Intertidal mudflats in low energy environment in An Bien, An Giang, VMD	The western side of VMD
3	Morphological Evolution of an Intertidal Mudflat in Relation to Mangrove Growth: Implications for Future Erosion Control (Nguyen et al. 2024)	Evolution of intertidal mudflat in Dong Thap, VMD	The eastern side of VMD
4	Intertidal Bare Mudflat and Wave Attenuation: A Case Study in the Vietnamese Mekong Delta (Loi et al. 2024)	Intertidal mudflats in wave dissipation and mangrove growth in Tran De, Can Tho City, VMD	The eastern side of VMD
5	Vegetated Intertidal Mudflat and Existing Ecological Restoration Perspectives to Control Coastal Erosion: Constraints and Recommendations (Nguyen 2025)	Evolution of intertidal mudflats along the VMD coast	The entire VMD coast
6	Is It Ecologically Feasible to Enclose Mangrove-Mud Coasts Using Multiple Concrete Structures for Erosion Control? Evidence from the Vietnamese Mekong Delta (Nguyen et al. 2025a)	Evolution of intertidal mudflats in estuaries in Dong Thap, VMD.	The eastern side of VMD

(continued)

Table 5.2 (continued)

No.	Paper/sources	Content	Remarks
	Natural growth of mangrove species in an eroding coastal area		
7	*Melaleuca* Entrapping Microsites as a Nature-Based Solution To Coastal Erosion: A Pilot Study in Kien Giang, Vietnam (Nguyen 2018)	Mangrove growth in eroding Vam Ray area, An Giang, VMD	The western side of VMD
8	Community Perspectives on an Internationally Funded Mangrove Restoration Project: Kien Giang Province, Vietnam (Nguyen et al. 2016)	Mangrove growth and sediment accumulation on eroding areas in An Giang, VMD	The western side of VMD

identifying sites of natural processes in the VMD coast (Table 5.2).

The results of the critical review provided basic from where open-ended questions were developed and then used in semi-structured interviews.

In the second step, 25 field visits were organized either by boat or by motorbike, to verify sites of natural processes and identify similar sites along the VMD coast. During these visits, the sites and their status were photographed and documented. Open-ended questions were first tested in an interview with a technical staff member working for Dong Thap DARD. The open-ended questions were then revised for subsequent semi-structured interviews with DARD's technical staff and group interviews with local residents living in and around these sites (Table 5.3).

Seven semi-structured interviews were conducted with two technical staff of each DARD at the DARD office of each province. Each interview lasted approximately one hour. Seven group interviews were conducted with the support of the seven DARDs of seven coastal provinces in VMD, with a total of 15 interviewees. Two criteria were used to select focus group members: (a) individuals who have lived in the area for a long time, (b) individuals who are involved in aquaculture fishing and development in these areas.

Table 5.3 The open-ended questions used in semi-structured interviews

Open-ended questions (English translation)
1. What did the area look like in the past?
2. When did intertidal mudflats evolve?
3. When did the mangrove species grow?
4. What have the community members done about these natural processes?
5. What lessons have you learned about these natural processes?

Purposive sampling (Neuman 2011) was used for these focus group interviews as participants with knowledge of mangrove forests, mangrove-based aquaculture, and muddy coasts were required for this study. Individuals were given the information about the study and interview, including their rights and the right to refuse the interviews. Verbal consent was obtained from individuals before the interviews were undertaken. Written consent or agreement was not common here as this was not traditional practice in the region. The data and information from field observations, photographs and semi-structured interviews were analyzed and categorized into diagram of drivers contributing to erosion using participatory diagramming (Neuman 2011). All the findings from detailed observations and semi-structured interviews were analyzed and synthesized into diagrams that show the evolution of intertidal mudflats and natural growth of mangrove species along the VMD (Fig. 5.1).

5.2 The Evolution of Intertidal Mudflats

During the field visits, six sites of natural processes were found along the VMD coast. These were Ngang Island (Dong Thap Province), Thanh Phu and An Thuy (Vinh Long Province), Ca Mau Cape (Ca Mau Province), and An Bien and Hon Dat of An Giang Province. Four of the sites were related to the evolution of intertidal mudflats. The remaining two were areas where mangrove species grew naturally in areas experiencing severe erosion. Due to insufficient chronological documentation (e.g., photographs,

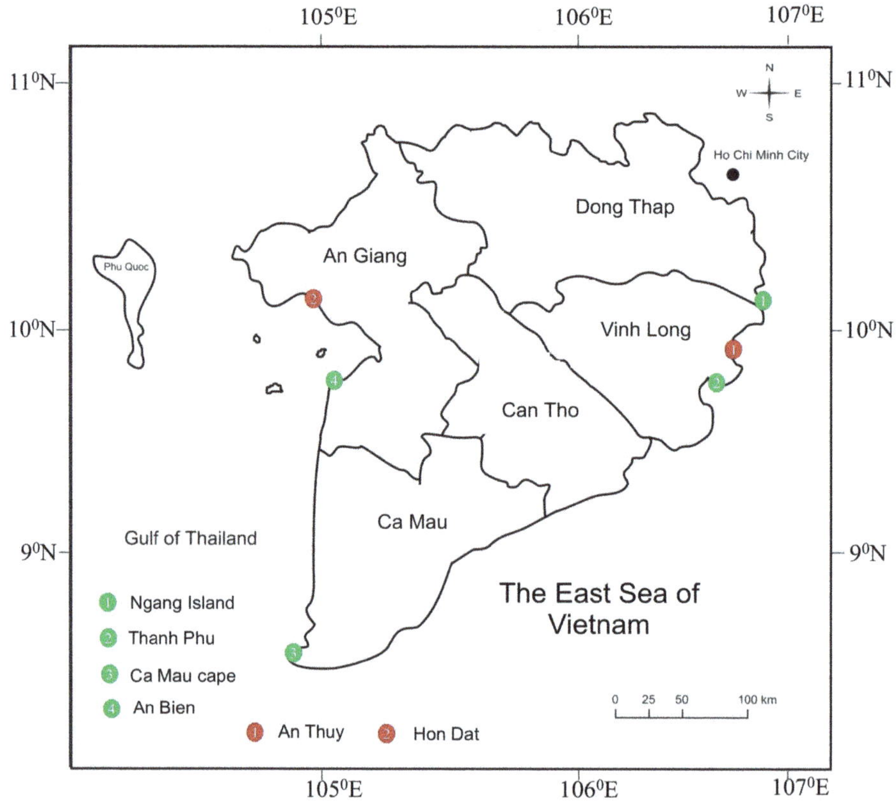

Fig. 5.1 Sites of the natural processes in the VMD coast. Green dots indicate the evolution of intertidal mudflats in low-energy environments, and red dots show natural growth of mangrove species in severe eroding areas. (The background map is Google image)

Table 5.4 Description of An Bien intertidal mudflat

Name/location/GPS recording	Site description	Management	The natural process
An Bien/An Giang province/9.830430° N; 104.958009° E	Intertidal mudflat established gradually over time. Located in river mouth areas (Cai Lon and Ca Be rivers) in enclosed bay. Approximately 420 ha (350 ha of mangrove vegetation cover and 70 ha of intertidal mudflat)	Part of Kien Giang coastal mangrove protection since 2008. No exploitation of coastal/mangrove resources.	Intertidal mudflats with increased sedimentation annually. Natural recruitment of mangrove species, mainly *Avicennia* species. Sediment accumulated gradually and compact over time.

progress reports, and measurement data) illustrating the evolution of the intertidal mudflats in Ca Mau Cape and Thanh Phu, these two sites were excluded from this chapter. However, this did not diminish the quality of this chapter as other areas exemplified the natural processes that occurred along the VMD coast.

5.2.1 The Evolution of An Bien Intertidal Mudflat in An Giang Province

An Bien intertidal mudflat is located within An Giang (formerly Kien Giang) Coastal Management Protection Forests. This area has

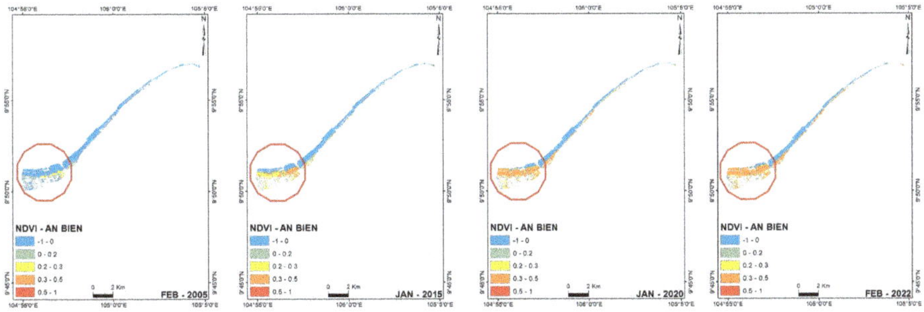

Fig. 5.2 The establishment of An Bien intertidal mudflat between 2005 and 2022 (red circle shows the location of An Bien intertidal mudflats). (Adapted from Nguyen and Nuong (2023))

Fig. 5.3 An Bien prograding area (**a**) is an intertidal mudflat (**b**) naturally growing *Avicennia* trees

been increasingly prograding over time since 2005 (Table 5.4 and refer to Fig. 5.1 for the location of An Bien intertidal mudflat).

Nguyen and Nuong (2023) revealed that fine-grained sediment accumulated over time as did vegetation cover between 2005 and 2022. The area is located in an enclosed bay and at the estuaries of Cai Lon and Ca Be rivers. Levels of vegetation cover increased steadily, while levels of non-vegetation cover decreased accordingly (Fig. 5.2).

An Bien intertidal mudflat prograded significantly between 2013 and 2016. This area was not allocated for local livelihood improvement. This area was usually protected and regularly patrolled by forest rangers of the management board of Kien Giang (now An Giang) Coastal Mangrove Protection Forest (Luom et al. 2021). Neither clearance nor interventions were permitted in the area.

The field surveys indicated that the area the area has been prograding over time due to sediment trapped in the area over time. Seeds of mangrove species were found trapped in the area. Mother trees, mainly *Avicennia* species were also found in the area. *Avicennia* species was found to have been the dominating species growing naturally in the area. The larger the intertidal mudflats

Table 5.5 Description of Ngang Island, Dong Thap province

Name/location/GPS recording	Site description	Management	The natural process
Ngang Island, Dong Thap province/GPS location: 10°12′45.29″N; 106°46′49.78″E	An island 800 m off the shoreline River mouths (Tieu and Dai rivers) in enclosed bay Approximately 910 ha (including 485 ha of intertidal mudflats and 425 ha of vegetated area)	Ngang island has been increasingly accreted as an island since 2005.	Intertidal mudflats with increased sedimentation gradually since 2005 Sediment accumulation with coarse-grained sediment (mainly sand) Natural recruitment of *Avicennia* species since 2010

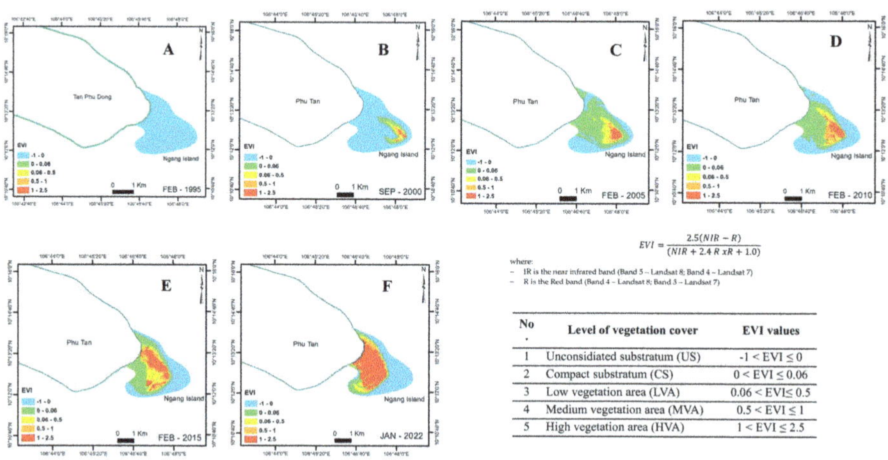

$$EVI = \frac{2.5(NIR - R)}{(NIR + 2.4\,R\,xR + 1.0)}$$

where:
- IR is the near infrared band (Band 5 – Landsat 8; Band 4 – Landsat 7)
- R is the Red band (Band 4 – Landsat 8; Band 3 – Landsat 7)

No.	Level of vegetation cover	EVI values
1	Unconsidiated substratum (US)	$-1 < EVI \leq 0$
2	Compact substratum (CS)	$0 < EVI \leq 0.06$
3	Low vegetation area (LVA)	$0.06 < EVI \leq 0.5$
4	Medium vegetation area (MVA)	$0.5 < EVI \leq 1$
5	High vegetation area (HVA)	$1 < EVI \leq 2.5$

Fig. 5.4 The morphological evolution of Ngang island in Dong Thap province between 1995 and 2022 (Adapted from Nguyen et al. (2024))

increased in size, the faster Avicennia trees grew faster over time to cover the area with mangrove trees (Fig. 5.3).

5.2.2 The Evolution of Ngang Intertidal Mudflat in Tan Phu Dong, Dong Thap

Ngang Island is located off the Phu Tan coast, Dong Thap (formerly Tien Giang) province, VMD. This area is part of the Dong Thap Coastal Mangrove Protection Forest. This area has been a protected area since 2005 (Tien Giang PPC 2013) (Table 5.5 and refer to Fig. 5.2 for the location of An Bien intertidal mudflat).

Nguyen et al. (2024) showed that Ngang Island has been increasing prograding over time. Sediment, fine-grained and coarse-grained, accumulated significantly over time. This gradual sediment accumulation provided a favorable condition for mangrove trees to grow, especially for seeds that floated in the area due to strong incident waves and tides (Nguyen et al. 2024, 2025a, b) (Fig. 5.4).

Field surveys indicated that Ngang Island is located in the estuaries of the Tieu and Dai rivers. Over time, sediments transported from these rivers have accumulated in this area. The island's location in a low-energy coastal environment made this gradual accumulation of sediment possible. This sediment accumulation created unconsolidated substrates that were seasonally

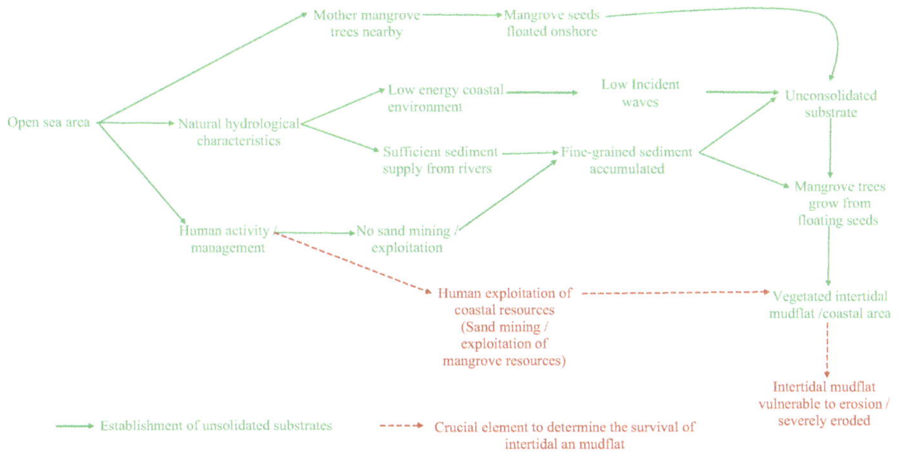

Fig. 5.5 Summary of the establishment of intertidal mudflats in prograding coastal areas and key elements to their survival. (Adapted from Nguyen et al. (2025a, b))

inundated by tides, providing favorable conditions for floating mangrove seeds to become trapped and grow naturally in the area. The growth of mangrove species then compacted and elevated these substrates. Between 1995 and 2022, the area of unconsolidated substrates decreased dramatically, whereas the area of vegetation increased. Areas where the mangrove trees grew were more compact than areas with newly accumulated sediment.

5.2.3 Summary of the Evolution of an Intertidal Mudflat in Natural Environments

Analyzing these two case studies produced a diagram showing where intertidal mudflats occurred in natural environments. These mudflats formed in areas with favorable hydrological characteristics, such as low incident waves and a sufficient supply of fine-grained sediment from river discharge. Mangrove seeds also floated onshore from nearby mangrove trees. Over time, the accumulation of fine-grained sediment resulted in the formation of an unconsolidated substrate. These substrates trapped floating mangrove seeds and provided favorable conditions for mangrove trees to grow naturally from the trapped seeds. The development of vegetated intertidal mudflats was driven by natural mangrove growth, continued sediment accumulation, and the establishment of unconsolidated substrates. However, the process was not complete until human exploitation of coastal resources was excluded. Figure 5.5 summarizes the entire process.

5.3 The Natural Growth of Mangrove Species in Actively Eroding Areas

5.3.1 Natural Growth of Mangrove Species in Vam Ray, Hon Dat, An Giang

An area of natural mangrove growth, mainly comprising *Avicennia* species, was found along the coast of Vam Ray in Hon Dat, An Giang Province, which experienced severe erosion between 2006 and 2008. Previous attempts to stabilize the erosion through the planting of mangrove seedlings and the construction of sea dykes failed due to an insufficient understanding of the area's morphology and dynamics (Nguyen et al. 2017) (Table 5.6).

Table 5.6 Description of Vam Ray site

Name/location/GPS recording	Site description	Human interventions	Natural mangrove growth
Hon Dat/An Giang province/10.171284° N; 104.816120° E	The entire Vam Ray area was severely eroded in 2008. Mangroves have been gradually destroyed. The area's mangroves remain geographically and ecologically fragmented, dominated by *Avicennia marina*, *Rhizophora apiculata*, *Bruguiera cylindrica*, *Bruguiera gymnorhiza*, and *Nypa fruticans*.	The area has been part of Hon Dat coastal mangrove protection forest. Planting of mangrove (*Rhizophora* and *Avicennia* species). The area was controlled through the construction of fences and revetments in the area. The planting failed as *Eucalyptus* pole fence was not effective in protecting the seedlings from strong waves. Many seedlings were uprooted by strong waves, resulting in a high death rate.	Floating mangrove seeds (mainly Avicennia species) were trapped in fragmented/ eroded areas. Seeds came from mother mangrove trees in the area or floated onshore from nearby mother trees. Mangrove seeds were trapped either by aerial roots of mangrove trees or marine debris washed onshore by strong incident waves.

Fig. 5.6 Mangrove seeds trapped in the areas with protection provided by marine debris and standing mangrove trees in the area. (**a**) and (**b**) show two severe eroding sections of Vam Ray with large burrows underneath. (**c**) and (**e**) show natural growth of *Rhizophora* trees. (**d**) shows natural growth of *Avicennia* trees

Field surveys and the findings by Nguyen (2018) indicated that despite the severe erosion process, seeds of mangrove species (mainly *Avicennia* and *Rhizophora* species) had become trapped in marine debris, such as dead and decaying mangrove trees and plastic bags, and the aerial roots of standing mangrove trees in the area. This debris provided a shelter for the seeds (Fig. 5.6).

Table 5.7 Description of An Thuy site, Vinh Long province

Name/location/GPS recording	Site description	Human interventions
An Thuy/Vinh Long Start point: BT-9°58′25.76″N; 106°38′34.94″E End point: BT-9°58′15.30″N; 106°38′14.58″E	The area was severely eroded with mature mangrove trees being uprooted. Sand accumulation buried aerial roots of mangrove (*Avicennia*) trees, killing most of the trees in the area *Grasses* grew naturally on sand accumulation areas. Trapping mangrove seeds drifting ashore and mangrove trees grew.	The area has been protected as coastal mangrove protection forests.

Fig. 5.7 Mangrove seeds trapped in the areas with protection provided by marine debris and standing mangrove trees in the area. (**a**)–(**d**) show seeds of *Avicennia* species growing in An Thuy eroding area

5.3.2 Natural Growth of Mangrove Species in Eroding Coastal Area in An Thuy, Vinh Long

The natural growth of mangrove species (mainly *Avicennia* species) was observed in many eroding coastal sections of An Thuy, Vinh Long (formerly Ben Tre) province. Eroding areas were covered by sand. Previous attempts to control erosion included planting of *Casuarina* trees and seeds or seedlings of *Avicennia* species (Table 5.7).

Field surveys indicated that despite severe erosion process, seeds of mangrove species (mainly *Avicennia* and *Rhizophora* species) had been trapped in marine debris (remnants of dead mangrove trees, decaying trees, weeds, and plastic bags) as well as the aerial roots of standing mangrove trees in the area. This marine debris provided a shelter for seeds to grow (Fig. 5.7).

5.4 Conclusions

This chapter summarizes the natural processes recorded along the VMD coast. The natural processes included the establishment of intertidal mudflats in natural coastal environments and natural growth of mangrove species in areas experiencing severe erosion. The establishment of intertidal mudflats depended heavily on sediment supply, marine hydrological elements, and the absence of human activity. By contrast, the natural mangrove growth depended significantly on existing physical conditions, such as seed sources, marine debris, and the exclusion of human activities.

References

Loi LT, Nguyen TP, Nguyen LT (2024) Intertidal bare mudflat and wave attenuation: a case study in the Vietnamese Mekong Delta. Ecol Eng 206:107320. https://doi.org/10.1016/j.ecoleng.2024.107320

Luom TT, Nguyen TP, Smithers S, Tai TV (2021) Protected mangrove forests and aquaculture development for livelihoods. Ocean Coast Manag 205:105553. https://doi.org/10.1016/j.ocecoaman.2021.105553

Neuman WL (2011) Social research methods: qualitative and quantitative approaches, 7th edn. University of Wisconsin, Boston

Nguyen TP (2018) *Melaleuca* entrapping microsites as a nature based solution to coastal erosion: a pilot study in Kien Giang, Vietnam. Ocean Coast Manag 155:98–103. https://doi.org/10.1016/j.ocecoaman.2018.02.005

Nguyen TP (2025) Vegetated intertidal mudflat and existing ecological restoration perspectives to control coastal erosion: constraints and recommendations. Ecol Eng 215:107610. https://doi.org/10.1016/j.ecoleng.2025.107610

Nguyen TP, Nuong CT (2023) Thinning and selective harvesting and mangrove protection forests: lessons learned and recommendations from the Vietnamese Mekong Delta. Estuar Coast Shelf Sci 288:108345. https://doi.org/10.1016/j.ecss.2023.108345

Nguyen TP, Tam NV, Quoi LP, Parnell KE (2016) Community perspectives on an internationally funded mangrove restoration project: Kien Giang Province, Vietnam. Ocean Coast Manag 119:146–154. https://doi.org/10.1016/j.ocecoaman.2015.10.008

Nguyen TP, Cottrell A, Parnell KE (2017) Human activities and coastal erosion on the Kien Giang coast, Vietnam. J Coast Conserv 21(6):967–979. https://doi.org/10.1007/s11852-017-0566-9

Nguyen TP, Thuan NB, Loi LT, Quoc HV (2024) Morphological evolution of an intertidal mudflat in relation to mangrove growth: implications for future erosion control. Life 14(6):711. https://doi.org/10.3390/life14060711

Nguyen TP, Thuan NB, Tien NN (2025a) Is it ecologically feasible to enclose mangrove-mud coasts using multiple concrete structures for erosion control? Evidence from the Vietnamese Mekong Delta. Ecol Eng 219:107678. https://doi.org/10.1016/j.ecoleng.2025.107678

Nguyen TP, Thuan NB, Tien NN (2025b) Monitoring the effectiveness of emergent detached offshore structures in mangrove vegetation increase: lessons and recommendations. Life 15:136. https://doi.org/10.3390/life15020136

Perricone V, Mutalipassi M, Mele A, Buono M, Vicinanza D, Contestabile P (2023) Nature-based and bioinspired solutions for coastal protection: an overview among key ecosystems and a promising pathway for new functional and sustainable designs. ICES J Mar Sci 80(5):1218–1239. https://doi.org/10.1093/icesjms/fsad080

Tien Giang PPC (2013) Decision to approve forest protection and planning in Tien Giang between 2011 and 2020. Tien Giang PPC, Tien Giang

UNEP (2020) The economics of nature-based solutions: current status and future priorities. UNEP, Nairobi

Wu L, Ouyang Y, Cai L, Dai J, Wu Y (2023) Ecological restoration approaches for degraded muddy coasts: recommendations and practice. Ecol Indic 149:110182. https://doi.org/10.1016/j.ecolind.2023.110182

Yasmeen A, Pumijumnong N, Arunrat N, Punwong P, Sereenonchai S, Chareonwong U (2024) Nature-based solutions for coastal erosion protection in a changing climate: a cutting-edge analysis of contexts and prospects of the muddy coasts. Estuar Coast Shelf Sci 98:108632. https://doi.org/10.1016/j.ecss.2024.108632

6 NbS to Control Severe Erosion in Kien Giang, VMD

Abstract

This chapter summarizes the development of NbS for protecting eroding Vam Ray coast from further erosion. The Vam Ray community members developed these solutions using their own knowledge and with technical support from external sources. These solutions included the production of high-quality mangrove seedlings mimicking mangrove ecological characteristics, the pilot construction of entrapping sites made of *Melaleuca* small-sized sticks mimicking natural trapping of seeds/propagules of mangrove species, and the gradual expansion of mangrove areas through the construction of *Melaleuca* fences among fragmented mangrove patches to protect planted seedlings from erosion and expand the mangrove area towards the sea. The effectiveness of these solutions has been evidenced through scientific publications.

Keywords

Coastal erosion · Fragmented mangroves · Local knowledge · Melaleuca poles

6.1 Introduction

Local ecological knowledge is defined as a cumulative body or system of knowledge and understanding about environmental factors and ecological dynamics, which is shared by local resource users (Davis and Wagner 2003). This type of knowledge has three attributes: (a) a system of practical knowledge and understanding of the environment shared by a group of resource users; (b) the knowledge accumulated through direct experience in a specific physical setting; and (c) the knowledge that is passed down through generations, usually through oral tradition (Davis and Ruddle 2010).

Local ecological knowledge has been increasingly used for coastal and fisheries management; for example, it has been used to manage marine conservation and customary fishing in southern Maharashtra state, India (Karnad 2022), to improve local fisheries management and conservation in Viana do Castelo, northern Portugal (Gomes et al. 2025), to understand a living coastal resource in Oregon, West Coast of North America (Sjostrom et al. 2021), and to manage marine protected areas and coastal fishing in Algeria (Boubekri et al. 2022). It is increasingly recommended for developing coastal erosion control measures in many areas around the world.

Vam Ray is located 50 km northwest of Rach Gia, An Giang (formerly Kien Giang) province. The Vam Ray coast (428 m) was eroded in 2005

P. T. Nguyen, *Nature-based Solutions in Mangrove-mud Coast Protection*, Sustainable Development Goals Series, https://doi.org/10.1007/978-981-95-9941-7_6

and became severely eroded in 2009. Earth dykes in the area were breached between 1992 and 2009, and mature *Rhizophora* trees were uprooted by seasonal strong waves, especially in high tides. This left vulnerable mangrove areas containing *Bruguiera cylindrica*, *Nypa fruticans*, *Rhizophora apiculata*, *Avicennia marina*, and *Sonneratia alba*. The erosion destroyed local aquaculture ponds, killed planted mangrove seedlings, and damaged local roads in Vam Ray area.

Between 2000 and 2008, DARD invested substantially in upgrading the existing earthen dykes and restoring the eroding Vam Ray area. These restoration measures included planting of seedlings of *Avicennia* species, which were nursed for at least 6 months in local nurseries and the establishment of offshore fences made of *Eucalyptus* poles to protect the seedlings. Despite the efforts failing, they were repeated annually. Technical staff of DARD were responsible for the technical designing and implementation, and the Vam Ray community was not involved in the entire process. Some Vam Ray community members were contracted to plant seedlings, protect and maintain the measures.

The Vam Ray community has lived in this area for decades. They have been involved in protecting mangrove forests and using allocated mangrove forests for aquaculture development for local livelihood improvement. Some senior residents developed an understanding of the area's natural characteristics as well as insights gained from past failures. However, these insights were rarely documented and were often passed down to younger generations during tea and coffee breaks or community festivals. Their knowledge was completely disregarded during the design and construction processes for two reasons. Firstly, local government agencies and professionally trained staff are responsible for developing coastal protection measures, with support from central government agencies. Secondly, insufficient documentation of the Vam Ray community knowledge had limited the Vam Ray community to share it with technical staff and government agencies when opportunities arose. Consequently, local involvement was limited to providing daily labor, supplying local materials, and assisting in monitoring infrastructures/structures constructed on sites.

From 2009 to 2011, the Vam Ray community developed NbS with technical and financial support from the Integrated Conservation and Development of Kien Giang Biosphere Reserve Project. This project was funded by the Australian Agency for International Development and implemented by the German Agency for Development Cooperation. A series of workshops and technical training sessions were organized in Kien Giang (now An Giang) province, along with field visits to other sites in Vietnam, to improve understanding of existing coastal protection measures. Vam Ray community members participated in these events. Subsequently, several meetings were organized with Vam Ray community representatives to review and document their knowledge related to coastal erosion control through participatory diagramming, photo sharing, and field observations. This process occurred alongside the introduction of scientific knowledge about the underlying causes of coastal erosion and the erosion process itself by consultants and researchers who were contracted by the project. The integration of local and scientific knowledge resulted in the development of specific coastal erosion control measures for the Vam Ray area.

All relevant parties agreed that DARD was responsible for constructing sea dykes and sluice gates and Vam Ray community, while the Vam Ray community and the project were in charge of developing coastal erosion control by combining the community's knowledge with scientific expertise. The project facilitated the integration process. Lessons learned from the Vam Ray case study have been widely reported in the literature (GTZ 2012a, b, c, d; Cuong et al. 2015; Nguyen et al. 2016a, b; Nguyen et al. 2017; Nguyen and Parnell 2017; Nguyen 2018). Therefore, this chapter summarizes the processes by which the Vam Ray community's local knowledge was documented, integrated, and used to develop coastal protection measures. The summary was compiled based on the author's personal experience of working as a technical officer on the project between 2009 and 2011, alongside the findings from group meetings in the Vam Ray community, the detailed documentation and observations from field visits conducted between 2013 and 2016, and the existing relevant literature, which was sufficiently cited where possible in the chapter. The processes were illustrated by the photographs taken by the authors and the visitors during field visits, which were adequateley cited in the chapter.

6.2 The Local Ecological Knowledge and NbS to Control Erosion

6.2.1 Production of High-Quality Mangrove Seedlings Mimicking Mangrove Ecological Characteristics of Mangrove Species (adapted from Nguyen et al. 2016a, b)

- *The Context*

In accordance with standard nursery procedures, seeds were collected and taken to nurseries where they were propagated in black plastic pots filled with a soil-compost mixture. The potted seeds were regularly irrigated with fresh water and nurtured for at least 6 months until they reached a height of at least 0.6 m. In some areas, the seedlings were not planted until they reached a height of 1 m. Seeds of *Avicennia* species were prioritized for collection and production because *Avicennia* is a pioneer species that helps accumulate sediment and grows on intertidal mudflats.

The most common problems encountered when planting included (a) low survival rates due to seedlings being uprooted by strong incident waves before their roots could take hold in the sediment, (b) seedlings dying from salt stress in salty planting environments, and (c) seedlings dying when their roots were cut prior to planting. Seedlings were nurtured in nurseries for a considerable amount of time, during which their roots grew into the soil through the pots. The roots were cut before the seedlings were transported to planting sites.

Standard plastic pots were usually sourced from big cities, which made them expensive. This posed particular difficulties for the Vam Ray community, given the condition of rural roads and low incomes (see Chap. 1 for details on sea dykes and rural roads).

The seeds/propagules of mangrove species were irrigated with freshwater in nurseries established in the surrounding area. However, this nursery technique caused stress to the seedlings when they were planted in coastal environments with saltwater, liquid sediment, and strong seasonal waves.

A few seeds were found in Vam Ray between May and July, which was an ideal time for planting. However, an abundant supply of of seeds from different mangrove species was available in the surrounding areas.

- *The Vam Ray Community's Knowledge of Mangrove Ecology*

The Vam Ray community members knew that *N. fruticans* would thrive in full sun, whereas *A. marina*, *R. apiculata*, *B. cylindrical*, and *S. alba* would fare better in partial shade. Seedlings were uprooted by strong waves and stressed in saltwater environments. Meanwhile, sources of seeds were not available locally at the time of planting.

Meanwhile, used plastic bags were found in the area. This was a crucial lesson learned at the site. Furthermore, a few seeds were found in Vam Ray between May and July, which was an ideal time for planting. However, an abundant source of seeds from different mangrove species was found in the surrounding areas.

- *The Vam Ray Community's Ideas for the Production of Seedlings Mimicking the Mangrove's Ecology*

The Vam Ray community came up with the idea of using used plastic bags for potting seeds and propagules both to help save money and reduce waste in the coastal environments. Plastic bags were perforated before potting. During the implementation, additional new plastic bags were sourced to ensure sufficient quantities. Seeds were irrigated with saltwater to help them adapt well to future planting sites (coastal and marine environments). Nurseries were constructed near the designated planting sites, and seedlings were regularly rotated to prevent their roots growing into the soil. Production of the seedlings started with the potting of three propagules of *Rhizophora* species, five propagules of *Bruguiera* species, one seed of *Nypa Fruitican* species and one seed of *Avicennia* species. More than three propagules of *Rhizophora and Bruguiera* species were used because these species required a strong foundation to protect them against being uprooted. This increased survival rates and facilitated rapid re-colonization if successful. Seeds of *Nypa Fruitican* species were planted so that local community members could collect leaves and fruits once they were fully grown, and it was planted near sea dyke toes to help stabilize the area with its roots. *Avicennia* species was often prioritized for planting. However, a small number of seeds and propagules developed roots that quickly penetrated the nursery ground despite being regularly rotated (Table 6.1; Figs. 6.1 and 6.2).

6.2.2 Construction of Entrapping Sites Made of Melaleuca Sticks Mimicking Natural Trapping of Seeds/Propagules of Mangrove Species (adapted from Nguyen 2017)

- *The Context*

Sediment accumulation in eroding coastal areas was a significant challenge. In restoration project, seedlings were planted with support provided by bamboo or *Eucalyptus* poles and concrete poles. However, the survival rate was low because seedlings were uprooted and washed ashore by strong seasonal waves and currents.

There was little demand for small *Melaleuca* sticks in local markets. Consequently, these sticks were often discarded or used as firewood. Previous attempts to control erosion in Vam Ray involved the use of small-sized *Melaleuca* sticks to support and protect planted seedlings. These sticks were left on site after the previous attempts failed.

Table 6.1 Summary of nursery and seedling production

Overall aim: Production of cost-effective seedlings using ecological elements of mangrove forests (sea mud for potting, irrigation using saline water to help seedlings adapt well to ecologically similar environmental conditions of planting sites, propagation of several seeds/propagule into clusters to improve survival against strong waves and to facilitate rapid re-colonization).			
Nursery construction	Seed collection	Potting	Nursery operation
- A nursery was established near intended planting sites in Vam Ray. Two-thirds of the nursery area were covered with fishing nets to shade mangrove seedlings from the sun - The nursery was constructed in May 2009 using *Melaleuca* poles and fishing nets for shading seeds	- Seeds of five different mangrove species were collected from nearby areas and stored in shady areas - Seeds were covered with wet banana or palm leaves or stored in wet plastic bags during transportation - Seeds were transported early in the morning or late in the afternoon to avoid direct sun	- Used plastic bags were used as pots and holed for irrigation - Liquid sea mud was collected at low tide and stored in holed bags for drying - Sea mud was used for potting instead of compost soil - More than one individual seeds/ propagules were propagated together in an attempt to increase survival rates during transplantation (*R. apiculata* had three propagules per pot, *B. cylindrica* had five propagules per pot, while *A. marina*, *S. alba*, and *N. fruticans* were potted as single individuals)	- Sea water was used for irrigating potted seeds - Potted seeds were regularly rotated to avoid roots growing into the soil - Potted seeds were nursed for 3 months in the nursery - Seeds of *Avicennia* species were propagated on the surface of soil and then transplanted into pots

Fig. 6.1 Collection of sea mud for potting in Vam Ray area, An Giang (formerly Kien Giang) province. (**a**) collection of sea mud at low tide and ready for filling used plastic bags. (**b**) used plastic bags filled with sea mud were dried before being transported to the nursery. (**c**) filled plastic bags were transported using wooden, iron-framed cart. (**d**) filled plastic bags were place in the nursery

Fig. 6.2 Potting mangrove seeds in the nursery. (**a**) shows five propagules of *Bruguiera* species potted in one pot. (**b**) shows three propagules of *Rhizophora* species potted in one pot. (**c**) shows propagules of *Bruguiera* species grew one week after potting. (**d**) shows one seed of *Nypa fruiticans* growing in a pot

Fig. 6.3 Floating mangrove seeds/propagules were trapped by entrapping sites found in the area. (**a**) seed of *Avicennia* species was trapped by remnants of fishing nets. (**b**) seed of *Avicennia* species was trapped in a hole inside a mature *Avicennia* tree. (**c**) seeds of *Avicennia* species were trapped by marine debris (i.e., rubber foam, remnants of plants). (**d**) propagules of *Bruguiera* species were trapped by marine debris (i.e., remnants of decaying plants and plastic empty bottles)

The Vam Ray community wanted to find an alternative solution to ensure the successful planting of mangrove seedlings and the sediment accumulation. The Vam Ray community wanted to expand their mangrove areas for aquaculture development and harvesting.

- *The Vam Ray Community's Knowledge of Mangrove Ecology*

The Vam Ray community members observed that seeds and propagules floated ashore by seasonal currents and waves. They became trapped in marine debris and/or the aerial roots of mangrove species (i.e., *Avicennia* and *Bruguiera* species). Over time, these seeds and propagules grew stronger over time, even in areas prone to erosion where there was no protection provided by offshore and shoreline structures (Fig. 6.3). Nguyen (2017) called this process entrapping sites.

- *The Vam Ray Community's Ideas for Constructing Entrapping Sites Mimicking the Mangrove's Ecology*

The Vam Ray community members constructed two entrapment sites using locally found *Melaleuca* poles. The first type involved constructing frames containing grids. These frames protected planted mangrove seedlings from being uprooted by strong waves and currents. Each grid was large enough to contain a potted seedling properly (Fig. 6.4).

The second type was constructed using discarded *Melaleuca* poles on sites from previous unsuccessful attempts. An additional amount of small *Melaleuca* sticks was sourced from local shops to ensure there was an adequate quantity for construction. This type helped to trap floating seeds and sediments for natural growth (Figs. 6.5 and 6.6).

Fig. 6.4 *Melaleuca* support frames. (**a**) Construction of a *Melaleuca* support frame. (**b**) and (**c**) *Melaleuca* support frames were used during the planting of mangrove seedlings in Vam Ray

Fig. 6.5 A section of entrapping site made of *Melaleuca* discarded sticks laid down on the ground

Fig. 6.6 *Melaleuca* entrapping sites trapped sediment and seeds of *Avicennia* species for natural growth. (**a**) *Melaleuca* discarded sticks. (**b**) *Avicennia* trees grew in the area. (**c**) sediment trapped over time in the area

6.2.3 Construction of Melaleuca Fences Among Fragmented Mangrove Patches to Protect Planted Seedlings from Erosion (adapted from GTZ 2012a, b, c, d; Cuong et al. 2015; Nguyen et al. 2016a, b; Nguyen and Parnell 2017)

- *The Context*

Two issues were identified regarding erosion control: (a) how to dissipate strong waves and (b) how to maximize survival rate of planted mangrove seedlings. Shoreline and offshore structures were constructed to dissipate the energy of strong waves. While these structures were effective in this respect, they were less successful in accumulating sediment and trapping floating mangrove seeds for natural growth. Additionally, the survival rate was low because the seedlings were uprooted by strong incident waves.

The restoration area was managed by the community members under allocation contracts. Meanwhile, the community was committed to ensuring a high survival rate of planted seedlings in order to restore their eroding allocated mangrove areas. They needed mangrove forests to protect themselves and their properties from erosion and strong incident waves and to improve their livelihoods.

- *The Vam Ray Community's Knowledge of Mangrove Ecology*

Field observations revealed that the remaining fragmented mangrove patches in the area naturally protected mangrove species that had grown from seeds landward and/or in front of sea dikes and sluice gates. Remnants of the *Rhizophora*, *Avicennia*, and *Bruguiera* species were found in the Vam Ray area. Natural mangrove forests have been effective in dissipating the energy of strong incident waves thanks to their species diversity and growth density.

Fences were made from Melaleuca poles, bamboo mats and used fishing nets were effective in preventing riverbanks from erosion. Bamboo mats and fishing nets were soft and pliable enough to dissipate the energy of strong incident waves without breaking. *Melaleuca* poles, bamboo mats, and old fishing nets could easily be found locally.

- *The Vam Ray Community's Ideas for Constructing Fences to Protect Planted Seedlings and Promote Natural Growth of Mangrove Species Mimicking the Mangrove's Ecology*

Table 6.2 Summary of construction of *Melaleuca* fences

Overall aim: Ensuring high survival rate of planted seedlings (designing of fences using local materials, *Melaleuca* support fences, planting seawards, and use of fragmented mangrove patches as natural protection, facilitation of sediment trap, and dissipation of the energy of strong incident waves).			
		Planting techniques	
Fence design and construction	Design and functions of *Melaleuca* frames	Planting strategy	Deployment of Melaleuca fences
- Fences were designed with *Melaleuca* poles, fishing nets, and bamboo mats. *Melaleuca* branches were used to fill the gaps between two rows - Fishing nets and bamboo mats were soft and supple to dissipate the energy of strong incident waves - Two rows of *Melaleuca* and the filling provided additional force to dissipate the energy of strong incident waves - Lateral fences were constructed single and/or double rows of *Melaleuca* poles as the first phase to reduce wave reflection and close gaps within fragmented mangrove patches - As the second phase, shoreline single fences were constructed with two supporting frames, bamboo mats, fishing nets and the filling of *Melaleuca* branches to reduce the energy of strong incident waves and to protect planted seedlings	- *Melaleuca* frames functioned as traps for protecting potted seedlings from being uprooted by strong incident waves. These frames were constructed with small grids as big in size as potted seedlings to stabilize potted seedlings - *Melaleuca* frames for providing extra force to shoreline single fence to dissipate the energy of strong incident waves and protect planted seedlings	- Transplantation took place in four phases, starting with the gaps within fragmented mangrove patches to strengthen these mangrove areas - As a second phase, seedlings were planted seawards with the protection of standing mangrove patches and shoreline single line fences - Planting of *N. fruticans* and *R. apiculate*, *B. cylindrica* in the gaps and close to the sea dykes as the first stage, while *A. marina*, and *S. alba* were planted in the later stages - A total of nine treatments and one control treatment were constructed with protection provided by the fences	- Lateral fences were constructed in the gaps to close the gaps and to create a safe place for the planting of potted seedlings to ensure high survival rate - Other fences were constructed following the lateral fences to ensure survival rate of planted seedlings and to expand gradually seawards - The fences were as high as the king tides in the region (+1.3 m) to facilitate sediment transported onshore by strong waves

The first step is the construction of *Melaleuca* fences in fragmented mangrove patches. This provided additional protection for the planted mangrove seedlings in addition to the protection provided by existing fragmented mangrove patches. This would ensure high survival rates. The fences were constructed with two rows of poles and frames made of bamboo mats, old fishing nets, and *Melaleuca* brushes, with small sticks placed in the gaps between the rows to create a strong foundation that would dissipate the energy of strong waves. The poles and frames werepushed as deep as possible into the mud manually to ensure strength and dissipate wave energy. The entire process is summarized in Table 6.2 and Fig. 6.7.

Fig. 6.7 Melaleuca fences deployed in Vam Ray. (**a**) and (**c**) were lateral *Melaleuca* fences. (**b**) and (**d**) were single line fences parallel to the shore

6.3 Conclusions

This chapter presents the Vam Ray community's local ecological knowledge regarding mangrove ecology, as well as the area's hydrological and morphological characteristics. They expressed interest in sharing their knowledge during the design and construction of coastal erosion control measures. However, they were completely ignored throughout the entire process because controlling coastal erosion was the responsibility of professionally trained technical staff. Meanwhile, the community gained knowledge through their daily observations of natural processes occurring in their area over time. Nevertheless, without technical and financial support from the project, their knowledge was not transformed into NbS for controlling erosion. The effectiveness of these solutions has been confirmed by various siientific publications.

References

Boubekri I, Mazurek H, Djebar AB, Amara R (2022) Social-ecological dimensions of Marine Protected Areas and coastal fishing: how fishermen's local ecological knowledge can inform fisheries management at the future "Taza" MPA (Algeria, SW Mediterranean). Ocean Coast Manag 221:106121. https://doi.org/10.1016/j.ocecoaman.2022.106121

Cuong CV, Brown S, To HH, Hockings M (2015) Using Melaleuca fences as soft coastal engineering for mangrove restoration in Kien Giang, Vietnam. Ecol Eng 81:256–265. https://doi.org/10.1016/j.ecoleng.2015.04.031

Davis A, Ruddle K (2010) Constructing confidence: rational skepticism and systematic enquiry in local ecological knowledge. Ecol Appl 20(3):880–894. https://doi.org/10.1890/09-0422.1

Davis A, Wagner JR (2003) Who knows? On the importance of identifying "experts" when researching local ecological knowledge. Hum Ecol 3:463–489. https://doi.org/10.1023/A:1025075923297

Gomes MA, Alves CM, Neto R, Faria F, Troncoso J, Gomes PT (2025) Fishing beyond the surface: harnessing local ecological knowledge to improve local fisheries management and conservation. Mar Policy 179:106763. https://doi.org/10.1016/j.marpol.2025.106763

GTZ (2012a) Coastal rehabilitation and mangrove restoration using Melaleuca fences: practical experience from Kien Giang province. CDBRP, Rach Gia, Kien Giang

GTZ (2012b) Erosion management and mangrove restoration plan for the mainland coast of Kien Giang, Vietnam in 2012. CDBRP, Rach Gia, Kien Giang

GTZ (2012c) Shoreline and mangrove resource condition of the mainland coastline of Kien Giang province, Vietnam: key pressures and restoration requirements in 2012. CDBRP, Rach Gia, Kien Giang

GTZ (2012d) Effectiveness of melaleuca fences for mangrove restoration rehabilitation efforts in 2012. CDBRP, Rach Gia, Kien Giang

Karnad D (2022) Incorporating local ecological knowledge aids participatory mapping for marine conservation and customary fishing management. Mar Policy 135:104841. https://doi.org/10.1016/j.marpol.2021.104841

Nguyen TP (2018) *Melaleuca* entrapping microsites as a nature based solution to coastal erosion: a pilot study in Kien Giang, Vietnam. Ocean Coast Manag 155:98–103. https://doi.org/10.1016/j.ocecoaman.2018.02.005

Nguyen TP, Parnell KE (2017) Gradual expansion of mangrove areas as an ecological solution for stabilizing a severely eroded mangrove dominated muddy coast. Ecol Eng 107:239–243. https://doi.org/10.1016/j.ecoleng.2017.07.038

Nguyen TP, Tong VA, Quoi LP, Parnell KE (2016a) Mangrove restoration: establishment of a mangrove nursery on acid sulphate soils. J Trop For Sci 28(3):275–284. https://www.jstor.org/stable/43856532

Nguyen TP, Tam NV, Quoi LP, Parnell KE (2016b) Community perspectives on an internationally funded mangrove restoration project: Kien Giang province, Vietnam. Ocean Coast Manag 119:146–154. https://doi.org/10.1016/j.ocecoaman.2015.10.008

Nguyen TP, Cottrell A, Parnell KE (2017) Human activities and coastal erosion on the Kien Giang coast, Vietnam. J Coast Conserv 21(6):967–979. https://doi.org/10.1007/s11852-017-0566-9

Sjostrom AJC, Ciannelli L, Conway F, Wakefield WW (2021) Gathering local ecological knowledge to augment scientific and management understanding of a living coastal resource: the case of Oregon's nearshore groundfish trawl fishery. Mar Policy 131:104617. https://doi.org/10.1016/j.marpol.2021.104617

7 A Way Forward in NbS in Mangrove-Mud Coast Protection in VMD

Abstract

This chapter provides a critical overview of the current situation regarding the sustainable management and use of the VMD coast for socio-economic development, as well as the practical experiences of local communities and relevant stakeholders in the region. As a new approach to NbS for the sustainable management and use of the VMD coast, this chapter emphasizes the importance of local communities in developing NbS to control erosion along the coast, in both policy and practice. Local communities should play a central role in coastal erosion control, rather than merely providing labour. Furthermore, coordination between government agencies, scientists and local communities should be established to facilitate their contributions. In practice, existing shoreline and offshore measures should be revised to take the natural conditions of designated erosion control areas into account more effectively. Furthermore, natural processes including rivers and riverine areas, and intertidal mudflats must be protected and incorporated into coastal protection measures. Adequate implementation of these recommendations will effectively protect the VMD coast and promote its sustainable management and use for socio-economic development in the future.

Keywords

Coastal vulnerability · Human factors · Natural factors · Restoration sites

7.1 An Overview of the Current Situation and Lessons Learnt

Over the past several decades, VMD has made significant contributions to Vietnam's social and economic development. However, its coast was severely eroded in the 1980s. This erosion resulted in the loss of coastal land and livelihoods, as well as the fragmentation and loss of mangrove forests that protected the coast. Coastal erosion was caused by natural and human-induced factors. Human-induced factors further fragment coastal mangrove forests and exposed already fragmented areas to erosion, thereby accelerating the process. Natural factors triggered coastal erosion processes that broke down these vulnerable areas, while in other areas, mangrove trees continued to grow naturally with high levels of sediment accumulation.

Since 2009, GoV has passed multiple legislative documents such as national-level policies and decisions that aimed to sustainably manage VMD, especially better to adapt to negative impacts of climate change and rising sea levels. These legislative documents provide legal frameworks that are then translated into detailed

P. T. Nguyen, *Nature-based Solutions in Mangrove-mud Coast Protection*, Sustainable Development Goals Series, https://doi.org/10.1007/978-981-95-9941-7_7

workplans at the lower government levels. These frameworks promote the use of local morphological and hydrological characteristics such as elevated areas, sand dunes, existing structures, and coastal natural resources and dynamic processes as an integral part of coastal erosion measures. In some areas, resettlement of erosion-prone communities and realignment of existing shoreline structures (i.e., sea dykes and sluice gates) are also possible options.

In practice, shoreline structures such as sea dykes and sluice gates and revetments and offshore structures such as Geo-tubes and offshore breakwaters have been emphasized along the VMD coast. Shoreline structures were constructed as measures to stop areas from further erosion, while offshore breakwaters were constructed parallel to the shore as long-term solutions to prevent coastal erosion. The primary objective of these breakwaters was to dissipate the energy of strong waves and to allow sediment to accumulate so that mangrove species can grow naturally. This was possible by the widely held view that the energy of strong waves and currents accelerated coastal erosion (Nguyen 2022). Due to their effectiveness in dissipating the energy of strong incident waves and their continued presence at these sites, these breakwaters were rapidly replicated in other provinces in VMD. By December 2025, half of the Kien Giang coast (79 km of Kien Giang shoreline) had been enclosed by these breakwaters. Please refer to Chap. 4 of this book for further information regarding the design and construction of these structures.

Offshore breakwaters were originally designed and installed parallel to the shore to protect eroding sandy coasts from further erosion, and strong wave energy was the dominant cause of erosion. These breakwaters were then modified in terms of their shape, size, and design, with gaps being created between sections and circular holes being added in order to accommodate mangrove-mud coasts. Despite being used to protect a different type of coast with different underlying causes of erosion, the technical design and modelling of these structures have remained unchanged. While these breakwaters have been effective at dissipating the energy of strong waves, they have had limited success in accumulating sediment as well as promoting the natural growth of mangrove species (Luom et al. 2021). Similar results were observed with shoreline and offshore revetments. Mangrove afforestation using seeds and propagules was usually carried out in intertidal mudflats where mangrove species grew naturally over time, rendering planting unnecessary (Nguyen et al. 2025). However, planting seedlings with bamboo or *Melaleuca* support sticks (Chap. 4) was not as successful as planned in eroding areas because the seedlings could not withstand strong waves. The planting was then reinforced with offshore concrete breakwaters, which increased construction costs and labor time.

By contrast, a previous study revealed evident signs of robust mangrove growth and sediment accumulation, even in eroding areas along the VMD coast, such as Vam Ray and An Thuy. Mangrove species grew naturally on sediment that had accumulated over time. These species then grew strongly, consolidating and compacting the liquid sediment, as well as elevating the area's substrates. Furthermore, mangrove trees grew strongly from seeds or propagules that were trapped by marine debris in eroding areas (Nguyen 2018).

The Vam Ray community was one of the few groups able to develop NbS using their detailed observations and knowledge of the mangrove's ecology, as well as the morphological and hydrological processes taking place in their area. However, the Vam Ray community was completely overlooked during the design and construction of coastal erosion control measures as it was believed that the design and construction of these measures are the responsibility of professional technical staff and relevant government agencies. Additionally, their knowledge and field observations were rarely documented and only shared with other community members at events.

With the financial and technical support of the AusAID funded project, the Vam Ray community transformed their knowledge into concepts and ideas for NbS to control erosion. These solutions included (a) the production of high-quality mangrove seedlings that mimic mangrove ecological characteristics; (b) the pilot construction of entrapping sites made of small *Melaleuca* sticks that mimick the natural trapping of mangrove species' seeds/propagules; and (c) the gradual expansion of mangrove areas through the construction of *Melaleuca* fences among fragmented mangrove patches to protect the planted seedlings from erosion and expand the mangrove areas towards the sea. These nature-based concepts and ideas have been successful in controlling erosion in Vam Ray, as evidenced by various scientific publications.

In summary, a significant discrepancy was found between policy and practice with regard to the use of NbS, the roles of local communities and local knowledge. Although natural processes have been emphasized in policy, they have not been adequately considered when designing and constructing coastal protection measures. Various scientific publications have confirmed the importance of these natural processes as a crucial element of NbS, their effectiveness, as well as their potential for replication in the region. Furthermore, local communities have successfully controlled the erosion and promoted the natural growth of mangrove species by applying their knowledge of natural processes gained from the surround areas to develop NbS. However, this knowledge and its owners have been completely overlooked throughout the entire erosion control process, despite the Vam Ray community's results being widely reported.

7.2 Current Threats to the Survival of the VMD Coast

The survival of the VMD coast has been increasingly under threaten from natural and human-induced factors. The former included rising sea levels and the negative effects of climate change. The VMD region is at high risk of inundation. If the sea level rose by 100 cm, approximately 38.9% of the delta would be at risk of flooding. The provinces at the highest risk are Hau Giang (80.62%), Kien Giang (76.86%), and Ca Mau (57.69%) (Vietnam MonRE 2020). The latter included exploitation of natural resources for regional development purposes , such as the construction of hydropower dams upstream of the Mekong River in China, Laos, Thailand, and Cambodia, with more recently planned (Trung et al. 2020), legal and illegal sand mining (Yuen et al. 2024), land subsidence (Minderhoud et al. 2017, 2018, 2019), the improper use of mangrove forests for aquaculture development (Nguyen et al. 2017), and the uncontrolled use of existing or newly established intertidal mudflats for growing cash crops or for coastal development activities (Nguyen 2015).

Previous studies have shown that the construction of hydropower dams has trapped sediment, thereby reducing sediment transport to the sea (Kondolf et al. 2014; Manh et al. 2015; Li et al. 2017; Bussi et al. 2021). Consequently, coastal areas have not been properly replenished, thereby exacerbating the vulnerability of the coastline (Fan et al. 2004). Sand mining has been extensively exploited in the region. Each province of VMD permitted sand mining to supply materials for housing or road construction (Yuen et al. 2024). Additionally, illegal sand mining along the rivers has been increased and become uncontrolled in the region (Yuen et al. 2024). Excessive sand mining likely caused environmental consequences for the riverbed, such as deeply incised riverbeds, which resulted in riverbank and coastal erosion. It also reduced water levels, disrupted hydrological connectivity, and diminished floodplain inundation (Park 2024). Along with sediment traps created by hydropower dam construction, excessive legal and illegal sand mining led to a substantial sediment deficit, which was still insufficient to compensate for the loss of extracted bed aggregates (Kondolf et al. 2014; Manh et al. 2015; Jordan et al. 2017; Lau et al. 2024). Land subsidence was closely related to natural and human-induced factors. Human-induced factors included large-scale land use changes, such as increased agricultural production, population growth, and urbanization in the delta. These changes were likely to alter the hydrological system or increase loading of the delta surface (Minderhoud et al. 2017, 2018, 2019). These changes amplified natural subsidence processes and created new anthropogenic subsidence (Minderhoud et al. 2018). As for the use of mangrove forests for aquaculture development, members of local communities ignored the requirements of their allocation contracts by illegally encroaching on protected mangrove forests. Consequently, mangrove forests have been substantially fragmented and even exposed to erosion (Nguyen et al. 2017). Due to the need to improve local livelihoods and income, as well as scarce land resources, people living in coastal areas tended to use these mudflats for various purposes. These mudflats were frequently ploughed to create land on which cash crops, such as watermelon, cassava, guava, and winter melon were grown to generate extra income. They extracted groundwater in these areas to irrigate these crops. Locals built houses in which they lived while growing these crops. In some areas, restaurants were built to care for daily tourists. Operating these restaurants required extracting groundwater, discharging wastewater, and disposing of waste (Nguyen 2015). By December 2024, many of the mudflats have subsided and disappeared permanently.

7.3 A Way Forward in Developing NbS in Mangrove-Mud Coast Protection in VMD

In recognition of the potential impact of natural phenomena, such as sea level rise and climate change and human activities on the VMD coast, GoV recently introduced a long-term strategic plan, running. From 2021 to 2030 with a vision extending to 2050 (VPM 2022), the plan aims to encourage sustainable and environmentally friendly development in the VMD region. It emphasizes the development of people-centered policies and protection of cultural, social, and natural ecosystems and water resources, in order to better adapt to the predicted negative effects of climate change and rising sea levels. The coastal provinces in VMD (the seven provinces before July 1, 2025) have successfully translated the GoV's 2022 plan into their own 10-year socioeconomic development plans. These provinces were Tien Giang (VPM 2023a), Ben Tre (VPM 2023b), Tra Vinh (VPM 2023c), Soc Trang (VPM 2023d), Bac Lieu (VPM 2023e), Ca Mau (VPM 2023f), and Kien Giang (VPM 2023g). Additionally, GoV has emphasized protecting the VMD coast from erosion and the potential effects of rising sea levels and inundation in the future (VPM 2022).

In the context of global climate change, NbS has increasingly been emphasized as an effective strategy for protecting against coastal erosion (Morris et al. 2022; Temmerman et al. 2013; Yasmeen et al. 2024). Examples of successful NbS solutions for controlling erosion on mangrove-mud coasts included the use of soft or biodegradable materials, such as coconut logs (Morris et al. 2022), oyster reefs (Chowdhury et al. 2021), bamboo fences, and permeable brushwood dams (Winterwerp et al. 2020). In light of this, the existing socio-economic development plans should be revised to incorporate natural process, NbS development as well as the roles and knowledge of local communities in the environmental protection section. This will ensure strategic measures are in place for the sustainable management and use of the VMD coast for socio-economic development, particularly with regard to adapting to climate change and rising sea levels. While it is not possible to prevent sediment trapping by hydroelectric dams upstream of the Mekong river, specific recommendations are provided below to effectively protect the VMD coast and promote its sustainable use for socio-economic development in the future. This will eventually achieved the ultimate goal of environmental protection, as stipulated in provincial-level socio-economic development plans and the GoV's 2020 plan.

- Firstly, existing socio-economic development plans stipulate that local communities should be involved as labor providers. As discussed in Chap. 6, local communities have a thorough knowledge of the natural processes occurring in and around their homes, having lived there for years and being most affected by coastal erosion. External individuals, such as scientists and technical staff, have not spent enough time in these areas, and therefore have less knowledge of coastal morphology and dynamics than the local population. Consequently, the existing socio-economic plans should be amended to recognize the roles and knowledge of local communities when designing and implementing coastal erosion control measures. To improve the role of local communities, there should be smooth coordination among scientists, government agencies, and local communities. This coordination would integrate different knowledge systems to develop effective NbS.

- Secondly, the natural processes of designated restoration sites must be taken into consideration in developing coastal protection measures. In particular, mangrove afforestation in intertidal mudflats is unnecessary, as these environments are low-energy and enclosed and mangrove species will grow naturally. There are many examples in literature and practice that have address this issue. If planting is deemed necessary, sediment accumulation should be prioritized over the planting of mangrove seedlings. Sediment accumulation can result from a sufficient sediment budget (Nguyen et al. 2025). Mangrove trees grow in areas with high levels of sediment accumulation.
- Thirdly, structures should only be used as a last resort to control erosion. The underlying causes of coastal erosion should be first identified and eliminated. Subsequent steps depend substantially on a critical situational analysis. Construction of structures should also be considered a temporary solution. If necessary, the coastal morphology and dynamics of sites designated for erosion control should be modelled in order to identify sedimentation processes, such as gross depositional sediment or sediment transport patterns, before offshore breakwaters are designed and implemented. Fourthly, riverine areas and intertidal mudflats must be strictly protected in order to promote natural processes which are a vital component of NbS as well as used as an integral part of coastal protection measure. Sand mining in rivers and riverine areas should be also closely monitored to prevent further disruption to sediment budget. Lastly, uprooted mangrove trees or remaining vegetation covers in eroded areas should not be cleared, but rather used to create entrapping sites that help trap floating seeds and sediment, providing favourable condition for mangrove species to grow naturally.

7.4 Conclusions

This chapter analyzes the current situation including the existing measures and threats to the VMD coast in relation to the use of NbS for controlling coastal erosion. Local communities and local knowledge have been proven to be significant contributors to the development of NbS for controlling the erosion along the VMD coast. Local communities should not merely be viewed as labor providers. It is important to note that the coordination among local communities, scientists, and governmental agencies must be stipulated in the socio-economic development plans to facilitate these contributions. The natural conditions of designated sites must be adequately considered when designing and implementing coastal erosion control measures. Furthermore, natural processes including rivers and riverine areas, and intertidal mudflats must be protected and incorporated into coastal protection measures. If implemented smoothly, all of these will effectively protect the VMD coast and promote the sustainable management and use of the VMD coast for socio-economic development in the future.

References

Bussi G, Darby SE, Whitehead PG, Jin L, Dadson SJ, Voepel HE, Vasilopoulos G, Hackney CR, Hutton C, Berchoux T, Parsons DR, Nicholas A (2021) Impact of dams and climate change on suspended sediment flux to the Mekong Delta. Sci Total Environ 755:142468. https://doi.org/10.1016/j.scitotenv.2020.142468

Chowdhury MSN, La Peyre M, Coen LD, Morris RL, Luckenbach MW, Ysebaert T, Walles B, Smaal AC (2021) Ecological engineering with oysters enhances coastal resilience efforts. Ecol Eng 169:106320. https://doi.org/10.1016/j.ecoleng.2021.106320

Fan LC, Deng B, Korotaev V (2004) The coasts of China and issues of sea level rise. J Coast Res 43:36–49

Jordan C, Tiede J, Lojek O, Visscher J, Apel H, Nguyen HQ, Quang CNX, Schlurmann T (2017) Sand mining in the Mekong Delta revisited—current scales of local sediment deficits. Sci Rep 9:17823. https://doi.org/10.1038/s41598-019-53804-z

Kondolf GM, Rubin ZK, Minear JT (2014) Dams on the Mekong: cumulative sediment starvation. Water Resour Res 50:5158–5169. https://doi.org/10.1002/2013WR014651.Received

Lau RYS, Park E, Koh YQ, Tran DD, Kantoush SA, Binh DV, Loc HH (2024) Recurrence interval of riverbed sand mining hotspots in the Mekong Delta: potential indications of unsustainable replenishment rates. J Environ Manag 370:122435. https://doi.org/10.1016/j.jenvman.2024.122435

Li X, Liu JP, Saito Y, Lap NV (2017) Recent evolution of the Mekong Delta and the impacts of dams. Earth Sci Rev 175:1–17. https://doi.org/10.1016/j.earscirev.2017.10.008

Luom TT, Nguyen TP, Anh NT, Tung NT, Tu LX, Duong TA (2021) Using fine-grained sediment and wave attenuation as a new measure for evaluating the efficacy of offshore breakwaters in stabilizing an eroded muddy coast: insights from Ca Mau, the Mekong Delta of Vietnam. Sustainability 13(9):4798. https://doi.org/10.3390/su13094798

Manh PV, Dung NV, Hung NN, Kumma M, Merz B, Apel H (2015) Future sediment dynamics in the Mekong Delta floodplains: impacts of hydropower development, climate change and sea level rise. Glob Planet Chang 27:22–33. https://doi.org/10.1016/j.gloplacha.2015.01.001

Minderhoud PSJ, Erkens G, Pham VH, Bui VT, Erban L, Kooi H, Stouthamer E (2017) Impacts of 25 years of groundwater extraction on subsidence in the Mekong Delta, Vietnam. Environ Res Lett 12(6):064006. https://doi.org/10.1088/1748-9326/aa7146

Minderhoud PSJ, Coumou L, Erban LE, Middelkoop H, Stouthamer E, Addink LE (2018) The relation between land use and subsidence in the Vietnamese Mekong Delta. Sci Total Environ 634:715–726. https://doi.org/10.1016/j.scitotenv.2018.03.372

Minderhoud PSJ, Coumou L, Erkens G, Middelkoop H, Stouthamer E (2019) Mekong Delta much lower than previously assumed in sea-level rise impact assessments. Nat Commun 10:3847. https://doi.org/10.1038/s41467-019-11602-1

Morris RI, Bilkovic DM, Walles B, Strain EMA (2022) Nature-based coastal defence: developing the knowledge needed for wider implementation of living shorelines. Ecol Eng 185:106798. https://doi.org/10.1016/j.ecoleng.2022.106798

Nguyen TP (2015) Towards sustainable coastal mangrove protection and management in Ben Tre, Vietnam. IUCN, Gland

Nguyen TP (2018) *Melaleuca* entrapping microsites as a nature based solution to coastal erosion: a pilot study in Kien Giang, Vietnam. Ocean Coast Manag 155:98–103. https://doi.org/10.1016/j.ocecoaman.2018.02.005

Nguyen TP (2022) Offshore structures, sea mud accumulation and mangrove regeneration: insights from the Vietnamese Mekong Delta, Vietnam. Ocean Coast Manag 225:106243. https://doi.org/10.1016/j.ocecoaman.2022.106243

Nguyen TP, Cottrell A, Parnell KE (2017) Human activities and coastal erosion on the Kien Giang coast, Vietnam. J Coast Conserv 21(6):967–979. https://doi.org/10.1007/s11852-017-0566-9

Nguyen TP, Thuan NB, Tien NN (2025) Is it ecologically feasible to enclose mangrove-mud coasts using multiple concrete structures for erosion control? Evidence from the Vietnamese Mekong Delta. Ecol Eng 219:107678. https://doi.org/10.1016/j.ecoleng.2025.107678

Park E (2024) Sand mining in the Mekong Delta: extent and compounded impacts. Sci Total Environ 924:171620. https://doi.org/10.1016/j.scitotenv.2024.171620

Temmerman S, Meire P, Bouma TJ, Herman PMJ, Ysebaert T, De Vriend HJ (2013) Ecosystem-based coastal defence in the face of global change. Nature 504:79–83. https://doi.org/10.1038/nature12859

Trung LD, Duc NA, Nguyen LT, Thai TH, Khan A, Rautenstrauch K, Schmidt C (2020) Assessing cumulative impacts of the proposed Lower Mekong Basin hydropower cascade on the Mekong River floodplains and Delta – overview of integrated modeling methods and results. J Hydrol 581:122511. https://doi.org/10.1016/j.jhydrol.2018.01.029

Vietnam MOnRE (2020) Scenarios of climate change and sea level rise for Vietnam. Vietnam MOnRE, Hanoi

VPM (2022) Decision on approving master plan for the Vietnamese Mekong Delta from 2021–2030, with a vision up to 2050. VPM, Hanoi

VPM (2023a) Decision on approving master plan for Tien Giang from 2021 to 2023, with a vision up to 2050. VPM, Hanoi

VPM (2023b) Decision on approving master plan for Ben Tre from 2021 to 2023, with a vision up to 2050. VPM, Hanoi

VPM (2023c) Decision on approving master plan for Tra Vinh from 2021 to 2023, with a vision up to 2050. VPM, Hanoi

VPM (2023d) Decision on approving master plan for Soc Trang from 2021 to 2023, with a vision up to 2050. VPM, Hanoi

VPM (2023e) Decision on approving master plan for Bac Lieu from 2021 to 2023, with a vision up to 2050. VPM, Hanoi

VPM (2023f) Decision on approving master plan for Ca Mau from 2021 to 2023, with a vision up to 2050. VPM, Hanoi

VPM (2023g) Decision on approving master plan for Kien Giang from 2021 to 2023, with a vision up to 2050. VPM, Hanoi

Winterwerp JC, Albers T, Anthony EJ, Friess DA, Gijón Mancheño A, Moseley K, Muhari A, Naipal S, Noordermeer J, Oost A, Saengsupavanich C, Tas SAJ, Tonneijck FH, Wilms T, Van Bijsterveldt C, Van Eijk P, Van Lavieren E, Van Wesenbeeck BK (2020) Managing erosion of mangrove-mud coasts with permeable dams – lessons learned. Ecol Eng 158:106078. https://doi.org/10.1016/j.ecoleng.2020.106078

Yasmeen A, Pumijumnong N, Arunrat N, Punwong P, Sereenonchai S, Chareonwong U (2024) Nature-based solutions for coastal erosion protection in a changing climate: a cutting-edge analysis of contexts and prospects of the muddy coasts. Estuar Coast Shelf Sci 298:108632. https://doi.org/10.1016/j.ecss.2024.108632

Yuen KW, Park E, Dung TD, Loc HH, Feng L, Wang J, Charles-Robin G, Switzer AD (2024) Extent of illegal sand mining in the Mekong Delta. Commun Earth Environ 5:31. https://doi.org/10.1038/s43247-023-01161-1

GPSR Compliance
The European Union's (EU) General Product Safety Regulation (GPSR) is a set of rules that requires consumer products to be safe and our obligations to ensure this.

If you have any concerns about our products, you can contact us on

ProductSafety@springernature.com

In case Publisher is established outside the EU, the EU authorized representative is:

Springer Nature Customer Service Center GmbH
Europaplatz 3
69115 Heidelberg, Germany

www.ingramcontent.com/pod-product-compliance
Ingram Content Group UK Ltd.
Pitfield, Milton Keynes, MK11 3LW, UK
UKHW050301220926
475829UK00005B/109

* 9 7 8 9 8 1 9 5 9 9 4 0 0 *